Versandsc
Fahrzeugsteuerung
M-AT 4060

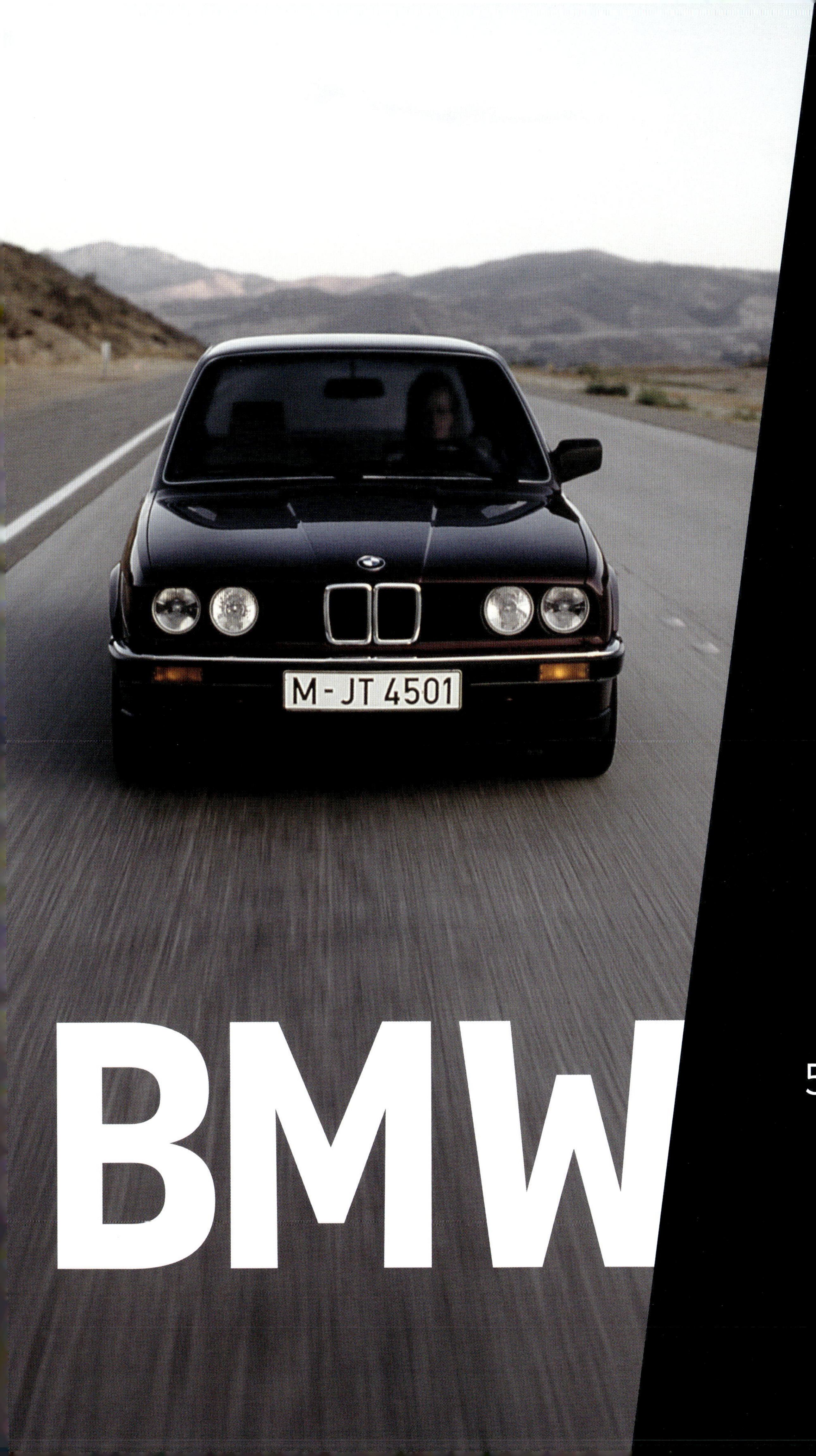

BMW

3-Series
50 Years

TONY LEWIN

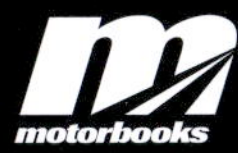

Quarto.com

First Published in 2025 by Motorbooks, an imprint of The Quarto Group,
100 Cummings Center, Suite 265-D, Beverly, MA 01915, USA.
T (978) 282-9590 F (978) 283-2742

EEA Representation, WTS Tax d.o.o.,
Žanova ulica 3, 4000 Kranj, Slovenia.
www.wts-tax.si

29 28 27 26 25   1 2 3 4 5

ISBN: 978-0-7603-9315-4

Digital edition published in 2025
eISBN: 978-0-7603-9316-1

Library of Congress Cataloging-in-Publication Data

Names: Lewin, Tony author
Title: BMW 3-series 50 years / Tony Lewin.
Other titles: BMW 3-series fifty years
Description: Beverly, MA : Motorbooks, an imprint of The Quarto Group, 2025. | Includes index. | Summary: "BMW 3-Series 50 Years offers a detailed history of this legendary model. Author Tony Lewin charts the 3-series' design and technical evolution from the 1970s to today"—Provided by publisher.
Identifiers: LCCN 2025012590 | ISBN 9780760393154 | ISBN 9780760393161 ebook
Subjects: LCSH: BMW 3 series automobiles—History | BMW 3 series automobiles—Design and construction | BMW automobiles—History | BMW automobiles—Design and construction
Classification: LCC TL215.B25 L494 2025 | DDC 629.222/2—dc23/eng/20250416
LC record available at https://lccn.loc.gov/2025012590

Design: Cindy Samargia Laun
Cover Images: James Mann
Endpapers: BMW AG PressClub and BMW Archive

Printed in Guangdong, China TT072025

**TO MY WONDERFUL WIFE AND LIFE-PARTNER CHRISTIAN.
SHE HAS BEEN BY MY SIDE FOR FIFTY YEARS—
JUST LIKE THE BMW 3-SERIES.**

M Power

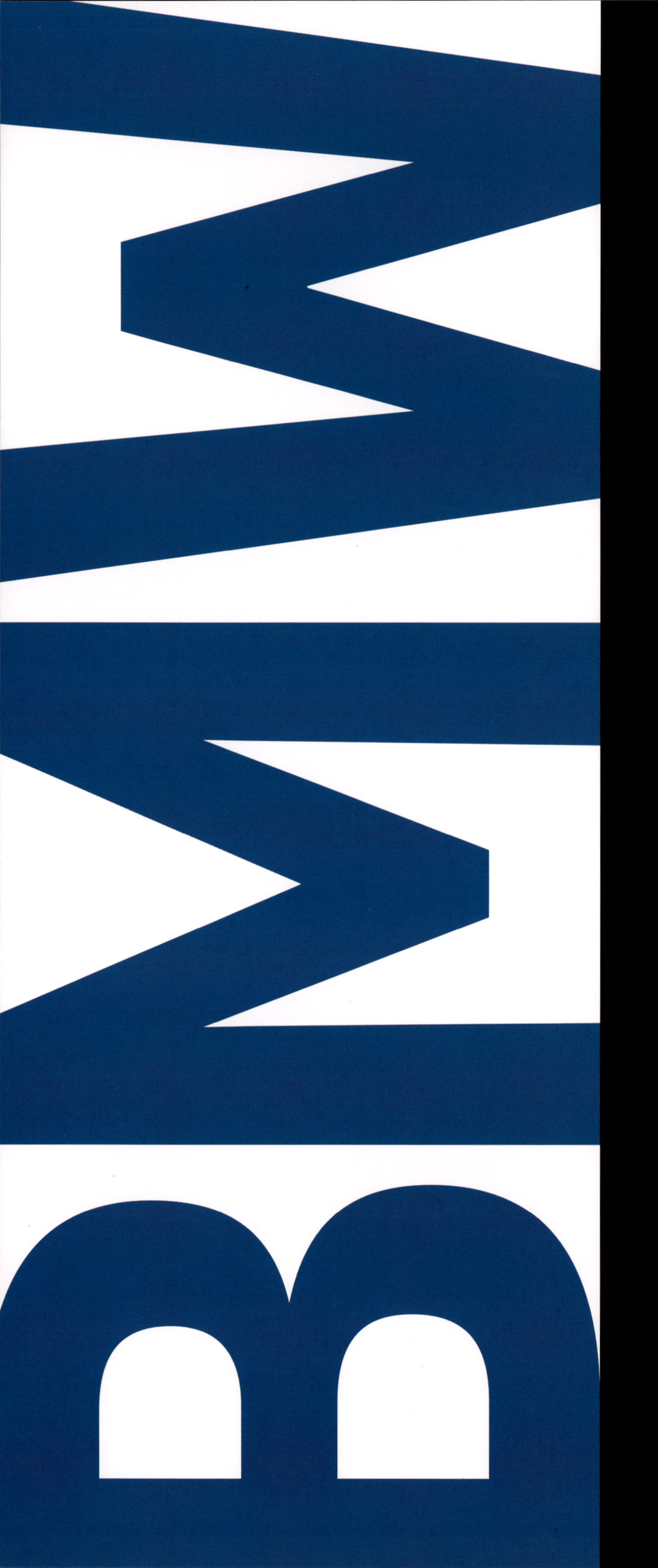

CONT

INTRODUCTION

The 3-Series *is* BMW. Nothing less. If there is a single most universally recognized BMW, it's the Three.

No doubt about it, the 3-Series is the emotional heart of the BMW brand, the most important individual model in the company's 110-year history. Most especially as it's the car whose pioneering sports sedan formula and spectacular success kicked off a massive upheaval in the car market and helped propel BMW to the leading position it enjoys today.

It did this by upending the stuffy postwar mindset that quality cars had to be big, pricey, and stodgy to drive; small cars were merely cheap and flimsy; and only sports cars could offer that magic ingredient—fun. But with the arrival of the 3-Series, drivers could suddenly engage with a classy quality car that was small, stylish, well put together, affordable and, vitally, sporty and fun to drive.

The recipe took off so rapidly that the 3-Series quickly became both the heartbeat and the engine of the company, the rolling expression of its engineering innovation and public persona. And the whole huge BMW organization has danced to its tune ever since.

To BMW internally, the Three is the hero model that put the company back firmly on its feet after the Neue Klasse and '02 series had rescued it from its near-drowning in the early '60s. To the petrol-head the Three has always been the car to admire, to aspire to, to make a song and dance about, and, above all, to enjoy.

For if there is a secret ingredient to every single one of the twenty-million-plus 3-Series built so far, it's that they're all great cars: fun to drive, lively, responsive, energetic, and with the unmistakable feeling that they were designed, developed, and built by like-minded people who truly understand what connects a car to the soul of a keen driver. What's even more miraculous is that this applies across the board, from the humblest 316 right through to the most extreme M3—even though the dopamine rush delivered by the latter might be orders of magnitude more intense. Like a stick of seaside rock, the fun message shines through clearly, no matter which way you slice it.

And that, perhaps more than anything else, is what has captivated all those millions of buyers, across all fifty years, seven generations, and scores of 3-Series permutations that have emerged from the Munich powerhouse. The design may have evolved and the engineering has raced ahead, but the central identity and the customer appeal have remained constant, and the core values of compactness, quality, and, crucially, those sporting genes continue to resonate worldwide.

It doesn't even matter that, ever since the launch of the X3 in 2004, BMW's massive expansion has been fueled by diversification into products further and further removed from its emotional sports sedan heartland. Those crossovers have risen to nudge out the 3-Series as BMW's top-selling nameplates in many markets, but they still carry the dynamic stamp of their mentor—and, whatever the sales figures say, the company's top managers continue to hold the Three strongest in their own affections.

None more so than right now, as I write in early 2025, with the impending arrival of the new Neue Klasse, the BMW for the electric age. This will be the next 3-Series, the eighth generation of the corporate icon and the biggest and boldest investment in the company's history. It is deeply symbolic that BMW has chosen this, its favorite son, to pour all of its innovatory energy into.

The chapters in this book chronicle how each successive generation has evolved, how each has been enticingly innovative but also comfortingly familiar—a finely judged but powerful combination that has worked outstandingly well for half a century. This time round, negotiating the bridge to a very different engineering philosophy, the stakes are even higher and BMW's balancing act between continuity and moving the game on will be put to its toughest test so far. But if anyone can pull that off, it's BMW.

D
M YT 2002H

BMW 3-SERIES PEDIGREE

The lineage that bred the global phenomenon which keeps on winning

Aero engines
1917–1938

R32 Boxer motorcycle
1923

Austin 7
1927

328 Roadster
1936

1916 1925 1927 1935

Dixi DA1
1927

3/15 Sports
1930

303
1933

327
1937

1939-1945: WW2"

motorcycle
1948
700 Cabriolet
1959
31 prototype
1950
700
1959
501 sedan
1952
1959 Crisis
1602-2002 Cabriolet
1967
1602-2002 Touring
1971
1602-2002
1966
1960
1970
Neue Klasse 1500-2000
1962
Turbo concept car
1972
2002 Turbo
1973
1980
3-Series E21
1975
E21 Cabriolet
E30 Cabriolet
1986
E30 Touring
1988
James Mann
3-Series E30
1982
1990
E30 M3
1986
Z1 Roadster

E36 Coupe
1991

3-Series E36
1990

E36 Touring
1994

James Mann

E46 Touring
1999

E46 Coupe & Cabriolet
1999

E46 Compact
2001

1-Series (E82)
2004

Touring (E91)
2005

James Mann

Coupe (E92) & Cabriolet (E93)
2006

3-Series F30
2011

201

1995 **2000** **2005** **2010**

James Mann

E36 M3
1993

E36 Compact
1994

Z3 Roadster (E36-7)
1995

3-Series E46
1998

E46 M3
2000

Z4 (E85)
2003

3-Series E90
2004

James Mann

M3 Coupe & M3 Cabriolet (E92/3)
2007

4-Series Coupe (F32)
2013

4-Series Cabri
2013

4-Series GranC
2013

M4 Coupe & Cabri
2013

Touring (G21)
2019

M3 Sedan (G80)
2021

M3 Touring (G81)
2022

iX3
2025

3-Series G20
2018

Neue Klasse
2026

2020

2025

4-Series Cabriolet (G23)
2020

i4 (G26)
2021

4-Series Coupe (G22)
2020

4-Series GranCoupe (G26)
2021

M4 Coupe & Cabriolet (G82/83)
2021

THE SPARK OF GENIUS

The 328 roadster revolutionized sports car racing when it won the prestigious 1936 Eifelrennen at the Nürburgring straight out of the crate. It quickly became the hero of the BMW brand and its

BMW's initiation into the car-building business is a well-known story. But for all that familiarity, it remains a remarkable and surprising sequence of events—to the extent that it's worth retelling here because it throws so much light on the engineering ethos and the energy that over the course of the succeeding decades would propel the company to the very top of the global automotive industry.

It was a cautious and inauspicious beginning that, at the time, would not have warranted newspaper headlines or made much of a ripple in the industry: A regional German motorcycle and onetime aero-engine builder moves in on a faltering local business that's landed a deal to build cheap cars under license from Austin of England, already one of the world's biggest automakers. Not exactly a star-studded start.

The year was 1928, just a year before the Wall Street crash, and market omens didn't look good. Big-name carmakers Mercedes and Benz had just combined forces to create the powerful Daimler-Benz colossus, Germany was in a state of near chaos under the fractured Weimar Republic, and the flimsy Austin Seven model that BMW directors had chosen was just one among scores of cut-price baby cars being touted by struggling start-up firms in a disorderly and declining internal market.

Worse, the Austin design itself had serious shortcomings, even though its Dixi-branded German incarnation had been selling steadily in the months before BMW moved in. Sir Herbert Austin was a flamboyant, high-profile British industrialist whose firm was best known as a builder and exporter of medium and larger cars. However, Sir Herbert secretly admired what Henry Ford had achieved in the US with the low-cost Model T, and since 1920 he had been working on his hunch that a very basic baby car could sell well—and get his debt-ridden company out of the financial scrape it found itself in.

Austin's board of directors opposed the idea vehemently, but Sir Herbert had his riposte ready: He simply bypassed the board completely, snatched a young draftsman from the drawing office, and set up shop in his own mansion, Lickey Grange. What would become the Seven was reputedly designed and drawn up on Sir Herbert's billiard table—and the vehicle that emerged in 1922, mere months later, was dramatically smaller and simpler in its specification than other four-seater cars on sale at the time.

By today's standards, many of its engineering solutions would be judged unwise shortcuts, especially in areas such as cooling, lubrication, bearings, and even braking—but it worked. The Seven sold well, and its success soon began to rattle Austin's archrival, William Morris, whose mainstay model was larger and more conservative. Before long, the two industry titans were engaged in a bitter price-cutting war that would permanently weaken both organizations and eventually lead to their unhappy forced merger as British Motor Corporation two decades later.

What this meant for BMW was that it had bought into a design built down to a very low price that fell far short of the high technical standards the German firm was accustomed to as a builder of quality motorcycles and high-performance, safety-critical aero engines. But the

ABOVE: Sir Herbert Austin in an early Austin 7 tourer. Austin pioneered simple, low-cost cars for the masses, but when BMW took up the design in 1927, it found serious flaws in safety and construction.

LEFT: BMW's experience with high-performance aero engines for the German military gave it a head start in the inter-war years. Here, test pilot Frank Zeno Diemer sets one of his many high-altitude records.

BMW's first foray into proper sports cars came in 1930 with the DA3 Wartburg, its tiny 748cc engine now boosted to 18 hp. This miniature two-seater was a strong performer on the racing scene and paved the way for the firm's future competition successes.

company's talented engineers and strategists immediately set to work to fix the DA 1's most obvious flaws, making the car stronger, safer, and more comfortable, as well as worthy of the BMW badge, which first graced its radiator in 1929.

QUALITY, INTEGRITY, AND A SPORTING SIDELINE

In quick succession, BMW's engineering team under Max Friz rolled out a series of upgrades that gave the DA 1, now relabeled 3/15, independent front suspension, proper integrated four-wheel brakes operated by a single foot pedal, and overhead valves for its four-cylinder engine, which nevertheless remained tiny at 748cc. By the end of the decade, the 3/15 had become a range of models spanning a variety of body types, one of which was a small two-seat roadster.

And it was that latter, toylike variant, with its rounded tail and cut-down doors, that opened the floodgates to a rush of sporting four-wheeled BMWs. The skills of the firm's motorcycle-racing experts transferred seamlessly to the car sphere and spurred a series of technical innovations that would culminate in the iconic 328 in 1936.

Tiny though it was, the 1930 "Wartburg" 3/15 sportster, with its higher compression engine now giving 18 horsepower, was a big success in competition throughout the decade, perhaps foreshadowing the equally compact 700 Coupe's successes in the late 1950s. But as far as true sporting prowess was concerned, the key developments came in 1933 when, as part of a major engineering overhaul, a new and much stiffer backbone chassis was introduced along with an extra pair of cylinders for the Austin-based engine. The six-cylinder result would become BMW's calling card, a key differentiator that would distinguish BMW from its competitors for the rest of the twentieth century—and well into the twenty-first.

NOW WE ARE SIX

It doesn't take a mathematician to realize that adding 50 percent to the cylinder count of a 788cc four-cylinder engine does not result in a huge, tire-smoking powerhouse. BMW's first six, in 1933, displaced just 1,182cc and gave 30 horsepower—enough to add immediate smoothness and sophistication to the 303 sedan that it was launched in. Those comfort-enhancing qualities might have been seen as useful sales attributes at the time, but the BMW engineers' sights lay much further ahead. Their true agenda was to lock in plenty of potential for future development.

Step by step, the six grew from 1.2 to 1.5 liters, and then on to almost 2.0 liters. At the same time, output rose from a

The BMW 315/1 sports car (center right) celebrates its world debut at the 1934 Berlin Auto Show. The lightweight two-seater's 1.5-liter six-cylinder engine gave 34 hp, and soon increased to 45 in the 1.9-liter 319/1.

That crucial step forward: BMW's first six-cylinder engine appeared in the 303 sedan of 1933. Displacing just 1.2 liters, it gave a smooth 30 hp: Three years later it had grown to 1.9 liters and 80-plus hp in the 328, and was winning races across Europe.

wheezy 30 horsepower to more than 55. But the real excitement was yet to come: In 1936, BMW presented the 328 roadster—the model that would come to encapsulate the brand even more powerfully than the six-cylinder engine or the double-kidney grille.

To say that the 328's debut at the Nürburgring in June 1936 was a sensation would be a serious understatement. Straight out of the box, this evocatively lithe, low, and smoothly futuristic two-seat roadster triumphed in its very first race, the famous Eifelrennen. At the wheel was BMW works motorcycle racer Ernst Henne, who trounced the opposition—including those in supercharged pure racing cars. This was a quantum leap in race-car performance, akin to that of the Lotus 49 on the late-1960s Grand Prix scene or the Audi Quattro's impact on the World Rally Championship in the 1980s. The 328 proved that brain can beat brawn, that David can be mightier than Goliath.

At the core of the 328's success was the very clever reworking of BMW's existing 2.0-liter engine, which had a new aluminum cylinder head housing efficient hemispherical

The 315/1 in action, showing its sleek lines, the soon-to-be-classic double kidney grille, and the cut-down doors that would become a feature of the legendary 328.

combustion chambers and an intricate pushrod and rocker mechanism to operate the splayed valves. The result, on triple carburetors, was an easy 80 horsepower in standard form and, with a safe rev limit well beyond 5,000 rpm, the potential for much more in competition trim.

Once customer-car delivery began in early 1937, the race wins clocked up quickly as privateer owners took their chances against bigger and heavier machinery. The 328 took over two hundred class and outright victories all over Europe, and a streamlined version finished fourth overall in the 1938 Le Mans 24-hour race. A streamlined version clad in a futuristic aerodynamic body by Italian coachbuilder Touring won the 1940 Mille Miglia with Fritz "Huschke" von Hanstein, later Porsche's competition manager, at the wheel, and BMW prototyped its own in-house Kamm Coupe streamliner, too. This was lost in the mêlée of the rapidly developing World War II, but BMW created a replica in 2010 to celebrate the seventieth anniversary of the Italian road race.

A SENSE OF STYLE

While BMW's top engineers were busy developing race-winning machinery and building the framework for an in-house competitions department, behind the scenes a proper design studio was taking form. Under the direction of Fritz Fiedler and, later, Wilhelm Meyerhuber, a strategy emphasizing aesthetic design as a separate discipline from body engineering emerged, and the results showed in the coherent range of models that followed on from the 303 of 1933. Not only was the 303 the first six-cylinder BMW, but it was also the first to properly introduce the trademark double-kidney grille—a feature that was later swept back and paired with faired-in headlights for an elegant, streamlined look.

As BMW cars became more stylish and more fashionable—especially the elegant 327 Coupe and its associated Sport Cabriolet—their popularity grew, and the single factory in Eisenach struggled to keep up with demand. In a dilemma that would repeat itself half a century later, BMW was faced with a key decision: Should it build a new factory to produce extra volume or move its existing products upmarket to boost revenue by increasing retail prices?

The choice must have seemed an obvious one at the time, given the company's high profile and keen engineering talent. Accordingly, the small four-cylinder cars were dropped, and work began on a new 3.5-liter, six-cylinder engine and two prestigious luxury models: the 335 four-door sedan and the four-seat cabriolet. Unfortunately, the all-consuming priorities of Germany's war effort stopped production of civilian machinery, with barely four hundred of these larger, Mercedes-rivaling limousines built. But a much more serious consequence of the switch to an overtly premium focus would reveal itself postwar, when BMW found itself without the right kinds of cars for very different market conditions.

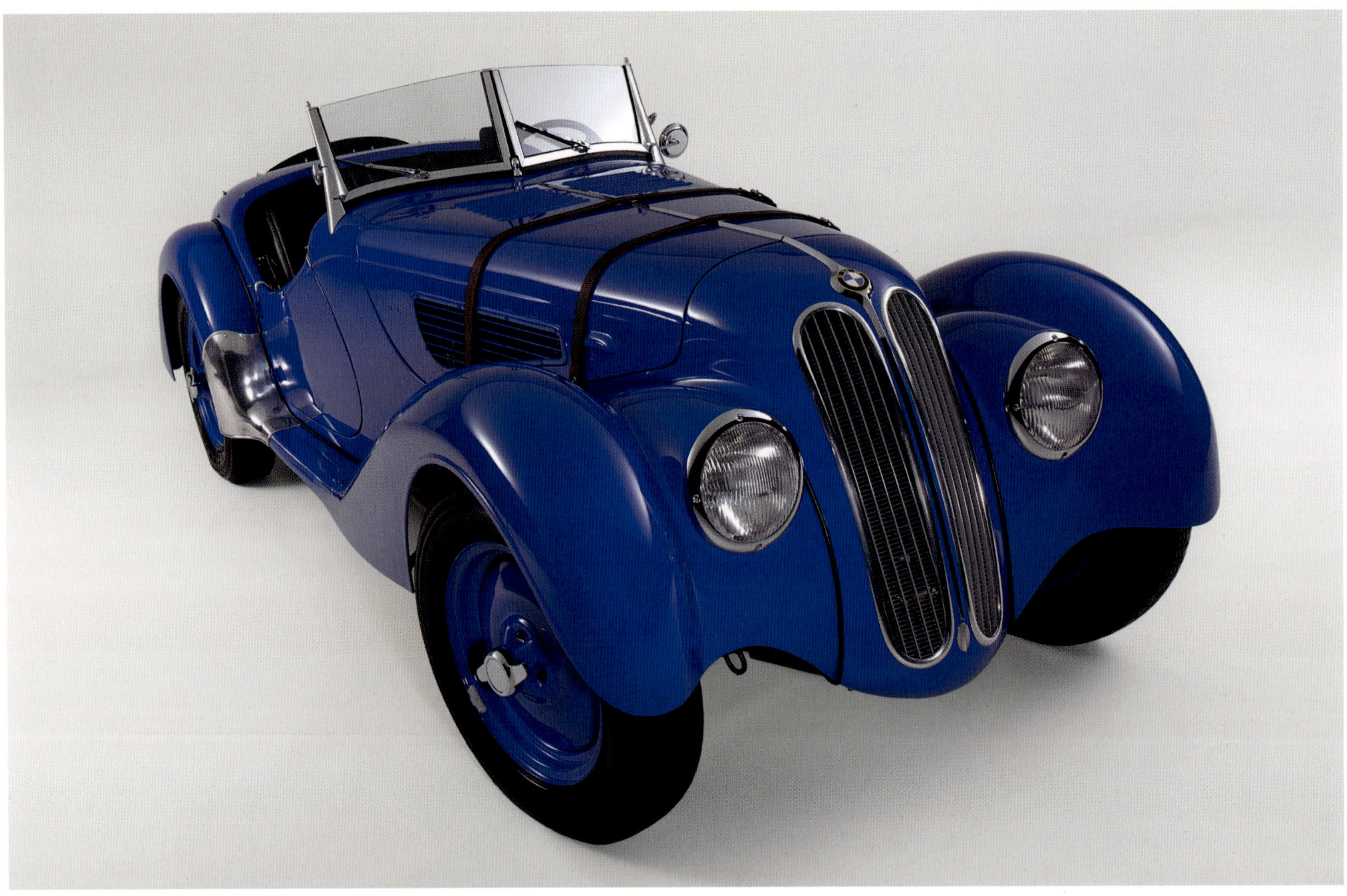

Innovation in action: the 328 sports car was a spectacular engineering breakthrough. Agile but powerful, this lightweight roadster triumphed in its first race against heavyweight opposition in the 1936 Nürburgring Eifelrennen.

STARTING OVER: IN SEARCH OF A NEW DIRECTION

It's no overstatement to say that by the time Germany surrendered to the Allies in May 1945, BMW had lost everything except its name. Its factories in southern Germany had been bombed to bits, the machinery destroyed or confiscated by the advancing American troops. Its Eisenach car plant—complete with design office, technical facilities, and equipment—was rendered inaccessible, as it lay in what was now the Soviet zone of occupation. Worse, the restrictions placed on German firms by the Allied powers forbade the manufacture of anything more advanced than pots and pans, bicycles, and agricultural implements.

Car production was out of the question, but it wasn't long before the manufacture of lightweight motorcycles was authorized. Soon, 250cc bikes followed the initial 125 model, and BMW staff began to return to the Milbertshofen plant in Munich. This in turn prompted managers and designers to rejoin. Interrupted only by the temporary setback of a quickly solved bankruptcy in 1948, company engineers began to secretly prepare for a return to car manufacture. For BMW, the inability to produce cars was particularly galling because the firm's prewar Eisenach plant had emerged from the war relatively undamaged, and the Soviets were exploiting their stroke of good luck by building BMWs there and selling them profitably into the Western Allies' zones.

BMW's problem was not simply the fact that cars from its former factory were being sold right under its nose; worse, the Munich remnant of the company had nothing with which to retaliate. Times were hard, and what limited demand existed in the market for cars was focused on simpler, smaller designs. The firm's four-cylinder models were gone, and all that was left in the cupboard was the prewar six—a situation that eventually provoked a significant high-level management split.

At the core of the argument was what sorts of cars BMW should now build. Overtures from BMW to license-build other automakers' models had failed, while chief engineer

Small-displacement motorcycles kept BMW going in the late 1940s, but they gradually began to fade as buyers sought the greater comfort and safety of small cars. The 250cc single from this bike would go on to great success in BMW's Isetta bubble car from 1955.

This smooth streamliner is actually the 1950 BMW 331 prototype, a two-seat sedan that could have been a big success but was canceled by BMW managers in a boardroom split. It used the well-proven flat twin motorcycle engine.

Alfred Böning had prototyped a sleek, compact two-seat coupe, much like the Fiat Topolino, that was powered by a flat-twin motorcycle engine. Styled by Peter Szymanowski, who had shaped many of the most admired cars that had built BMW's reputation before the war, the 331 had the makings of a major success—it could even have challenged Volkswagen's Beetle or Fiat's Nuova 500.

But sales director Hans Grewening had other ideas. With an eye on the much healthier profit margins promised by luxury cars, he killed the 331 project and instead commissioned the bulky 501 four-door limousine, again styled by Szymanowski in a swoopy, conservative idiom, and powered by the prewar six. Though the 501 had some interesting technical details, such as the center-mounted gearbox, it came across as bulbous and overbodied. It was expensive, too, and with just 65 horsepower on tap, its lethargic performance came as a deep disappointment to those with fond memories of BMW's agile and energetic prewar models. BMW would come to regret the decision to once again square up to Mercedes-Benz. The 501, unflatteringly dubbed the "Baroque Angel," sold poorly until it was later enlivened by a V-8 engine (the world's first all-aluminum unit of that configuration) with almost twice the power. That V-8, stretched to 3.2 liters and 120 or 140 horsepower, provided the chance for BMW to reenter the performance car arena, but again the opportunity misfired. The expensive and somewhat dowdy 503 coupe and convertible struggled to find buyers, while the ultra-glamorous 507 roadster stole all the headlines and became an unexpected image leader for the company. But the hard truth was that, iconic though it later became, the 507 was a resounding commercial failure and sold just 252 units in three years—even though it was endorsed at the time by famous buyers such as Elvis Presley and motorcycle race champion John Surtees. Today, surviving examples change hands for seven-figure price tags.

Wrong car, wrong time: Although the voluptuous 501 sedan is today revered for its styling excesses, it was a disappointing performer and too exclusive for the austere climate of the early 1950s.

The tiny bubble car that saved a big-name company: The BMW Isetta was licensed from Italy's Iso and used the 250cc motorcycle engine. It sold faster than any other car in the company's history and gave the firm breathing space to develop new models.

A BUBBLE CAR HELPS BALANCE THE BOOKS

While top managers at BMW remained hell-bent on challenging Mercedes in the luxury segment, those more in touch with reality were concerned about a slide in motorcycle sales as Germany's economic revival prompted bike buyers to trade up to the better safety and comfort of small cars. Motorcycles were BMW's only profitable activity, and a quick fix on the automotive side was essential to stave off complete collapse.

There had long been plans for a midrange 1.6-liter sedan, but the high costs of developing and tooling up for it meant it was an aspiration rather than a real possibility for the cash-strapped firm in this era. Instead, it was thanks to the smart thinking of BMW engineer Eberhard Wolff that a handy solution was found. On the Iso stand at the 1954 Turin motor show, Wolff spotted a tiny, egg-like car with a single door at the front, and a deal was rushed through allowing BMW to build the Isetta using BMW's own single-cylinder 250cc motorcycle engine. The little car was in production within months and rapidly became BMW's best-seller. Single-handedly, it kept the company afloat for a few more years, giving vital breathing space for engineers to develop the larger 600 "big bubble" four-seater and, critically, the dumpy 600's implausibly stylish spin-off, the 700.

ABOVE: BMW workers lower an Isetta bubble car body onto its chassis in the mid-1950s. Using the 250cc engine from BMW motorcycles, it was a big success with mobility-hungry buyers.

LEFT: BMW 700 sedans roll down the assembly line in the late 1950s. Fashionable and fun, it was the first BMW built with a modern integrated body and chassis.

Finally reaching outside the local Bavarian gene pool for design talent, BMW developed the 700 in a secretive skunk works in Vienna. Italian designer Giovanni Michelotti was already celebrated for the fashionable Triumph Herald and several Alfa Romeo, Lancia, and Ferrari studies. His elegant 700 was BMW's first properly modern car: Not only was its design chic, but underneath it was the first BMW to employ unitary monocoque construction rather than a separate chassis. It also carried over the innovative semitrailing arm rear suspension from the 600.

The rear-mounted, twin-cylinder 697cc motorcycle engine proved to be eminently tunable, the handling was excellent for the period, and BMW's board gave the project the green light. Such was the excitement around the 700 that orders came flooding in from far and wide, with more than twenty-five thousand clocked up in short order. It was an impressive debut, and it seemed that BMW had finally found its savior. Except that it was too late—or perilously near to being too late.

Developed in a secret Vienna skunk works by Giovanni Michelotti, the 700 blended trusted BMW engineering with eye-catching Italian style. Available as a sedan, coupe, or the convertible pictured, it was a huge success despite the company's shaky finances.

The world premiere of the BMW 1500 Neue Klasse at the 1961 Frankfurt show. Few debuts had been as excitedly awaited as this one, and with its fresh Italianate good looks and impressive technical specification, the orders soon began flooding in.

STARTING OVER—AGAIN

BMW was well accustomed to setbacks and must have been on first-name terms with its insolvency lawyers. But this crisis was worse than any before it. The crunch came in December 1959, when, at a rowdy shareholders' meeting, Daimler-Benz—which then owned a chunk of BMW—effectively pressed for the winding up of the company. To cut a complex story short, a loophole in the paperwork led to an adjournment and the miraculous emergence of a knight in shining armor, who promptly stepped in to snap up the surplus shares and keep the company out of Daimler's clutches.

As industrialists from a highly respected family consortium, Herbert Quandt and his brother Harald were immediately able to provide BMW with the financial stability that had eluded the company since 1945, allowing it to develop a long-term strategy and to put proper product planning and investment programs into place. Before long, many of the promising ideas that had bubbled beneath the company's surface since 1956 were dusted off and brought to the fore.

The Quandt family, which retains a controlling interest in BMW to this day, set the template for the astonishing success the company has enjoyed ever since. Key to Herbert Quandt's philosophy was providing the confidence and continuity to attract top talent, but staying clear of the company's day-to-day operations. A truly exceptional team soon fell into place to manage the company and develop the direly needed new car. Wilhelm Hermann Gieschen, Paul Hahnemann, Gerhard Wilcke, and, notably, racer turned engine wizard Alex von Falkenhausen were quickly on board. Styling would be attributed to Wilhelm Hofmeister, though his official role was head of body engineering, and the actual work was done by Italian studios Michelotti and Bertone as well as a rising internal talent, Georg Bertram.

A fleet of early Neue Klasse 1500 sedans about to go out to dealerships following the production start in 1962. The smart and impeccably engineered clean-sheet design went on to succeed way beyond BMW's wildest expectations.

The 1500's dashboard was as meticulously crafted as its exterior, setting new standards for presentation, quality, convenience, and comfort. This clarity has characterized every BMW since.

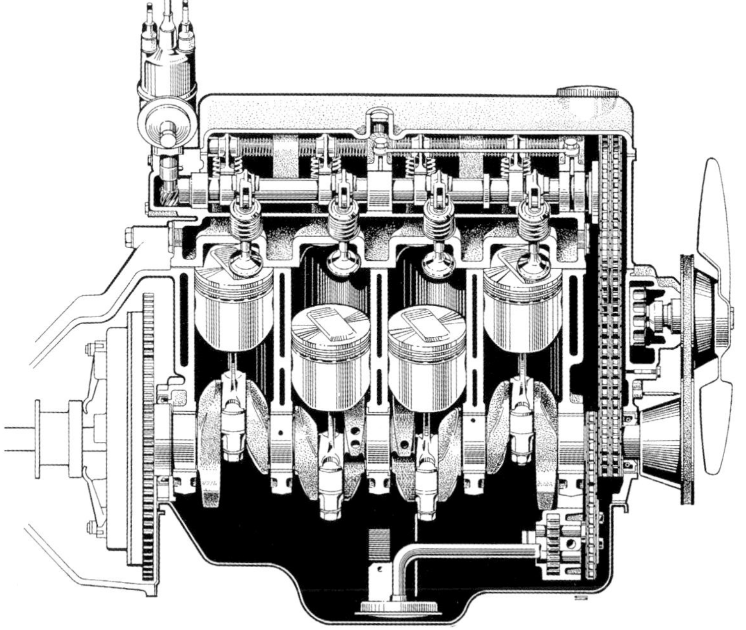

The overhead-camshaft M10 engine was central to the new car's success. Von Falkenhausen and his team deliberately designed in room for expansion from 1500cc—which soon paid dividends on the racetrack and in the marketplace.

The design that emerged in record time and bang on cue for the 1961 Frankfurt show stunned everyone. Its clean, crisp, and futuristic lines represented a breath of fresh air in an era of anonymous, rounded shapes; its light-filled interior sported an elegant, high-quality dashboard; and all-round independent suspension and disc brakes pointed toward great road manners. But it was von Falkenhausen's overhead-camshaft engine that grabbed the most headlines: Displacing just 1.5 liters, it boasted an unprecedented 75 horsepower, later increased to 80 for production versions, and promised eager performance in a midsize car weighing just 900 kilograms.

BMW had clearly got its mojo back, reconnecting with its sporting roots and finding genuine confidence for the first time in a generation. That confidence spilled over into the company's press communications for the 1500, already dubbed the first of the "Neue Klasse," or New Class, *Klasse* having the extra meaning in German of "classy" or "excellent." BMW's claims for the Neue Klasse were grandiose, to say the least, but the press absolutely loved it, Britain's *Motor Sport* praising its advanced technical specification, "luxuriously fitted out" interior, and "extremely large" trunk. If it could be sold for its equivalent German price of DM 8,500 in Britain, there would be no lack of buyers, predicted the normally conservative journal.

Aristocrat, designer, works car and motorcycle racer, and engineer extraordinaire, Alex von Falkenhausen was the inspiration behind the overhead-cam M10 engine that launched the 1500, won the drivers' Formula One world championship, and would stay in production for forty years.

And for once the hype wasn't just PR-department hot air. The 1500 was great to drive, marking it out as a much more enjoyable alternative to the stodgy Mercedes models that tended to be the default choice in the quality-car sector. *Autocar*, the world's oldest car magazine, gave the 1500 an enthusiastic welcome as an elegant five-seater that was "competitive on price"—in contrast to many of its continental counterparts—and described the performance figures of the 1800 model that soon followed as "quite outstanding." The only criticisms of note were inlet roar over 5,000 rpm, the tricky positioning of the ignition key, and the surprising fact that the model's electrics still adhered to an old-fashioned 6-volt system.

THE CORE PRODUCT LINE SPLITS IN TWO

So successful were the Neue Klasse cars that BMW's Munich factory was soon overwhelmed, and to keep up with demand, difficult decisions had to be taken. There was little protest when the Isetta was dropped from the range to make space for more 1500, 1600, and 1800 production, and few bemoaned the retirement of the brand's aging matriarch, the V-8 Baroque Angel—after all, barely eleven thousand had been sold over its twelve-year lifespan. But when the axe fell on the beloved little 700 coupe, questions were asked. Even so, the commercial logic was crystal clear. By 1965, the Neue Klasse, by now a three-model lineup ranging from the 1600 to the 1800 TI, was selling almost sixty thousand units a year, figures unheard of in BMW history and far beyond the Isetta at its popular peak.

In a textbook example of what would now be termed "premium pricing," BMW had intentionally positioned the Neue Klasse above humdrum contemporaries such as the VW 1500 and Opel Rekord. Its climb upmarket was underlined by the introduction of pricier and more powerful versions, culminating in the 110-horsepower 1800 TI. The perceptiveness of this policy, which effectively pioneered the premium market—as opposed to the luxury sector, which Mercedes-Benz monopolized—was emphasized by Eberhard von Kuenheim, who joined BMW in 1969 and would go on to lead the company for twenty-four years. In an interview with the author in 2004, he recalled how, in the early 1960s, the Quandts had made one of their rare interventions in BMW strategy when it came to pricing.

An early factory publicity shot for the Neue Klasse sedan, showing how BMW was learning the art of promoting the model to a new clientele focused on style and fashion.

The Neue Klasse's fine chassis, light weight and highly tunable engine soon led to victories in Europe's classic touring car races. The race-developed 1800 TISA boasted 130 hp and the later 2000 tii sedan, equipped with Kugelfischer fuel injection, set the template for future BMW sports and M models.

"The price of the Volkswagen Beetle in those days was between 4,000 and 5,000 marks," von Kuenheim explained. "Therefore, the board said they had to prepare for a very big crunch and introduce the Neue Klasse at 6,000 marks. Quandt, as an influential shareholder, dissented and told the board they were wrong, and that they should go for 8,000 marks—to show people it is a car with real value. It demonstrated that BMW was prepared to go against market research."

Von Kuenheim continued: "The higher price was well accepted. People said they could afford it and that they didn't want to drive a Beetle anymore—everybody has a Beetle. The effect of this was that BMWs became cars for '*Aufsteigers*,' the social climbers of the era, rising up the ladder of social status."

Intentionally, the premium shift left a gap at the lower end, waiting to be filled. Conveniently, BMW had just the car ready and waiting, in the shape of what was to become informally known as the '02 series. Though few could have foreseen it at the time, a decade later the '02 would, in its turn, serve as the precursor to an even more important car: the 3-Series.

The 3-Series is the core BMW model, remaining the company's top-selling nameplate for half a century and now spanning eight generations. More than any other single product, the 3-Series would come to define the BMW brand and its sporting ethos for countless millions of customers worldwide, proving instrumental in taking the company to the coveted top spot in the global premium market.

But that's jumping ahead too many generations at this stage of the game, for it is around the boardroom table in early 1965 that our 3-Series story really begins.

Ultimate development: The M10 engine enjoyed a parallel life on the racetrack in Touring Car racing and, finally in Formula One. Turbo boosted to a spectacular 1,200-plus horsepower in the Brabham-BMW BT52, it brought Nelson Piquet the world driver's title in 1983.

BMW 1500-2000 NEUE KLASSE: TIMELINE

1961	1500 four-door sedan unveiled at Frankfurt Motor Show, acclaimed for its fresh styling and engineering
1962	First production 1500s delivered, with engine boosted to 80 hp
1963	BMW 1800 launched, with 90-hp engine, servo brakes, and automatic option
1964	1.5-liter 1500 replaced by 1600, with more power, better equipment
1964	1800 becomes 1800 TI with twin carburetors and 110 hp; later, high-performance 1800 TI/SA added, with 130 hp, five-speed gearbox, sports seats, and race-developed chassis upgrades
1964	1800 TI race car finishes second at Spa 24 Hours in its debut season and goes on to win two further ETCC rounds in Zandvoort and Budapest
1965	Neue Klasse models collectively selling 60,000 a year—the highest figures in BMW history
1965	Lightened race 1800 TI/SA (160 hp) wins Spa 24 Hours
1966	1800 TI/SA and 2000 TI win 2.0-liter class in ETCC championship
1966	New 2000 model uses 100-hp engine from Coupe; TI version has twin carburetors and 120 hp
1966	1600 version dropped as two-door 1600-2 launched
1967	Tilux version added, with extra luxury equipment and restyled front with large rectangular headlights; facelift also including bigger rear lights, applied to other 2000s
1969	Fuel-injection 2000 tii first shown, the Kugelfischer system boosting power to 130 hp and top speed to 185 km/h
1971	All models begin phase-out as new 5-Series prepares for launch in 1972; total Neue Klasse sales exceed 350,000, BMW's best-ever result

M-HE 1495
M-PE 1421

COMPACT AND CAPTIVATING: LET THE FUN BEGIN

It may not carry the 3-Series family name, but the BMW 1602–2002 range is in almost every other respect the pushy parent and inspiration behind the real 3, which arrived nearly a decade later, in 1975.

If cars were movies, perhaps the '02 would now be seen as that crucial prequel to the upcoming big attraction, the preliminary foretaste giving a glimpse of the most important characters to follow. And of course this early preview quickly mushroomed into a phenomenon that led, whether by design or by sheer luck, to an astonishing sequence of ever more successful follow-ups—an auto franchise that over the next half century would reshape the whole industry and that, as it enters its critical eighth generation, is taking BMW into the brave new world of electric power.

And to labor the movie metaphor one last bit further, the '02 can be seen as having introduced the world to the script, the key central cast, and the leitmotifs that would propel the franchise forward for fifty-plus years: impeccable production values, carefully crafted plotlines, quick and agile twists and turns, and a strong force powering each successive production. The public quickly discovered a taste for these qualities, which had never been combined in a single package before, and—as evidenced by those twenty-million-plus customers who have so far flooded through the turnstiles—those deep-down values have never gone out of fashion.

HUMBLE BEGINNINGS

For a car that would turn out to be such an influential global phenomenon, the genesis of the '02 series was surprisingly prosaic. Way before executives began fussing about market segmentations or the price ladder between volume and premium products, the '02 arrived not so much as part of a clever grand plan but more as an opportunistic, low-risk gamble that happened to strike lucky.

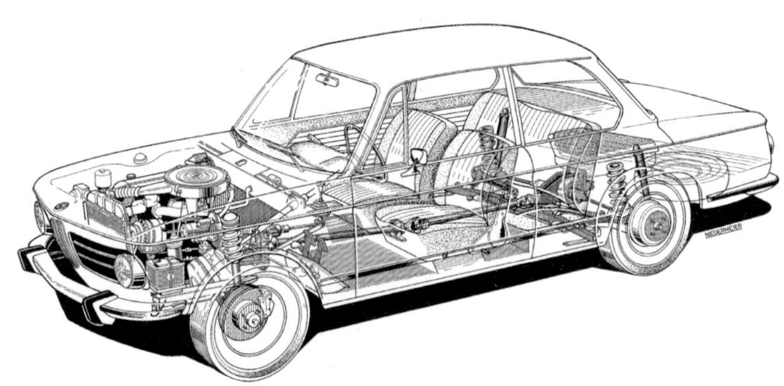

More than the sum of its parts: The '02 series was a simple engineering exercise, essentially a shortened Neue Klasse with two doors. Engines, transmissions, and suspension were all carried across.

Spring 1966, Munich Opera House square: BMW unveils the 1600-2 as a 50th birthday present to itself. Company strategists were wholly unprepared for its runaway success. Before long, the clumsy 1600-2 designation was simplified to 1602.

The '02 series was a simple formula, but the effect was magical. Its fun driving characteristics spoke to a new kind of buyer, one who valued agility and responsiveness rather than sheer power. Its instant success in the US put BMW on the map in North America.

BMW '02 SERIES: TIMELINE

1966	BMW 1600-2 launched as a two-door spin-off of Neue Klasse four-door sedan
1967	Twin-carburetor 1600 ti added, with 105 hp
1968	2.0-liter version added, with 100 hp; designation changed to 2002
1968	2002 gains second carburetor and 20 hp to become 2002 ti
1968	BMW wins European Touring Car Championship (ETCC) with 2002 ti
1969	Dieter Quester wins ETCC Championship in BMW 2002 ti with M121 Turbo engine with 240 km/h top speed
1970	Turbo version of 2002 ti wins ETCC Group 5
1971	2002 tii with 130-hp fuel-injection engine added
1971	Touring three-door hatchback version added
1971	Revised Cabriolet launched, with built-in roll bar; facelift across range, with bigger bumpers and rectangular rear lights; 1.8-liter 1802 version added
1973	2002 Turbo launched as world's first turbocharged car on general sale
1974	Turbo dropped after just 1,672 built
1975	Low-powered 1502 fuel-crisis special added; remaining models give way to new 3-Series
1977	1502 finally deleted, with 71,000 sold

BMW '02 SERIES: STYLES AND VERSIONS

Body Style	2-Door Sedan	2-Door Convertible	3-Door Touring
Length (mm)	4,230	4,230	4,230
Wheelbase (mm)	2,500	2,500	2,500
Curb Weight (kg)	940–990	1,040	1,030
Number Built	827,533	4,199	29,330

BMW '02 SERIES: BASICS AND ENGINEERING

Layout	Front engine, rear-wheel drive
Suspension F/R	MacPherson struts/semitrailing arms, coil springs
Steering	Worm and roller
Braking	Disc/drum

Too far ahead of its time: Hatchback-style Touring versions of the '02 series launched in 1971 with the promise of practicality as well as sporting style. But in a rare marketing failure by BMW, buyers stayed away and the model was dropped with fewer than thirty thousand sold.

It was an opportunity created by the success of the pioneering Neue Klasse, the fresh four-door sedans that had catapulted BMW from insolvency to international stardom in 1962. As the Neue Klasse's acclaim rose, so did its market positioning: Customers were forever demanding faster, more luxurious, and better-equipped versions, and the 1800 and 2.0-liter editions soon established themselves as the top sellers. This left a big opportunity lower down in the midsize car sector for a new entrant that offered the same BMW qualities of sportiness and agility but in a handier and more affordable format.

In engineering terms, the new model was a simple, low-stakes exercise. Effectively a shortened and simplified two-door version of the Neue Klasse, it capitalized on the donor car's well-proven and universally praised mechanicals. Even its external styling came across as wholly derivative of the larger car. That in itself was no bad thing, but on the face of it this seemed an inauspicious beginning for a model line.

MORE THAN THE SUM OF ITS PARTS

Launched in March 1966, the 1600-2 was quick to prove any doubters wrong. The clumsy 1600-2 designation was soon dumped in favor of the clearer 1602 (hence '02 series). Far from being a cut-down and stripped-out poor relation of the familiar four-door that some had feared, the 1602's smaller and lighter format served to intensify, rather than dampen, the BMW experience.

The parent four-door had been fêted as a smart, high-class family and business car that was also good to drive. But the 1602 was something else. Shorter, stubbier, and friendlier, its whole stance and attitude spelled alertness and agility. It came across as sporty and fun, even though it boasted only 85 horsepower and there weren't any remotely sporting design cues.

All this was enough to swing the balance sharply in favor of the 1602 and its much younger, more energetic personality. This was a car with real attitude, a cute coupe or hardtop to US eyes, but at the same time a sensible small car to suit European tastes; it spoke to a very new kind of customer. It was as if BMW had suddenly, and perhaps fortuitously, drilled into an untapped well of sentiment for its long-forgotten prewar sporting brand values. Reconnecting with the company's popular roots, the new model instantly raised BMW's game as far as public perception was concerned—especially in North America.

Early versions of the open-air '02 featured a full-length soft top but lacked rigidity. The 1971-on Targa-roof editions had a fixed roll hoop to provide strength and safety, sacrificing elegance in the process.

Simple, clear, and attractive: A well-designed driver environment has always been a strong selling point of compact BMWs.

Iconic designation: The three letters that came to mean so much to auto enthusiasts—and today they still resonate.

The sheer scale of the '02 series' success took BMW by surprise, and the brand suddenly found its products being promoted as fashion accessories for a glitzy city lifestyle.

The unmistakable rear view of an '02, with the length of the trunk perhaps exaggerated by the camera angle. The characteristic round rear lights lasted until 1971, when they were replaced by larger rectangular units.

RIDING THE ZEITGEIST

The effect was electric. Excitement among automotive commentators instantly infused the enthusiast community. The 1602's engine output was a modest 85 horsepower, but that went a long way in a 940-kilogram machine, and the little 1602 was unquestionably the car of the moment—the BMW everyone had been waiting for.

This explosive success is the stuff of history: how the '02 expanded into ever hotter and faster versions. How it hit a sweet spot in the US, becoming the favorite of *Road & Track* and other enthusiast journals—*Car and Driver* even named it the "world's best $2,500 automobile." BMW quickly exploited the car's high profile to cash in on customers' demands for more power and speed. That was simple: Just copy and paste all the technical upgrades introduced in the larger 1800 and 2000 sedans, and within two years the smaller car was outselling its bigger brother two to one.

As the '60s morphed into the '70s, there was still further excitement, and the fuel-injected 130-horsepower 2002 tii was being celebrated as the definitive most-fun sports car of the era. And this is when the seeds of the M-car bloodline first began to be sown.

THREE VARIATIONS, THREE MISFIRES

Along the way, the '02 models had spawned a couple of body variations. A neat Cabrio, outsourced to Baur, suffered from torsional flex and was soon replaced by a new version with a substantial roll bar that improved safety but detracted from the car's style. Neither open-top sold as strongly as expected, but the second body variation—the Touring—would come to be seen as the first actual mistake of BMW's post-1962 revival.

The late-'73 arrival of the 2002 Turbo stirred up a hornets' next of controversy, as much for its race car–like bodywork as for its sensational performance. This is an early example with the mirrored "turbo" script on the front air dam, a feature quietly dropped in response to complaints.

The logic behind the Touring model was irrefutable: Take a highly respected two-door compact car and give it the extra versatility of folding rear seats and a hatchback tailgate to allow unimpeded access to a large load area. There would be no dilution of the driver experience in either performance or handling, but a useful boost in practicality at no extra cost. The Touring name distanced it from station wagons and the fashionable hatchbacks that were then beginning to gain favor in the European sales charts.

To everyone's surprise, the Touring stuttered on the starting grid. Bafflingly, few buyers seemed to want it or need it, and the misstep would soon become a standard case study in the world's business schools. A favorite analysis put forward was that although the Touring might have been a very good BMW, it wasn't the kind of BMW people wanted—or even expected. First and foremost, BMWs were sport sedans, so they didn't need to be practical. And secondly, BMWs were conservatively shaped three-box designs, and any alteration to that silhouette detracted from their BMW-ness. Just 3 percent of the 850,000 '02 models built were Tourings, and, mindful of this embarrassment, BMW has ever since been cautious when evolving its shapes.

The third variation is much better documented. BMW was on something of a high in the summer of 1972 after the Olympic Games in its home city of Munich, and the dramatically futuristic Turbo concept car had gained lots of excited media coverage as it led the athletes' parade, drawing admiring stares from all who glimpsed it. The concept used a version of the 2.0-liter M10 engine, turbocharged up to the then-phenomenal output of 200 horsepower. Capitalizing on this star turn, BMW slotted the turbo engine, detuned slightly for road use, into what was basically a 2002 tii sedan.

Right car, wrong time: The 2002 Turbo was the world's first volume-produced turbo passenger car, but its debut coincided with the 1970s fuel crises and it was withdrawn after little more than a year. Just 1,672 were built.

TURBO TRAILBLAZER

The 2002 Turbo arrived in autumn 1973 as the world's first turbocharged passenger car, boasting 170 horsepower, a top speed of 211 km/h, and acceleration of 0–100 km/h in well under 8 seconds. These were pretty explosive figures for the time, but what outraged people more than the performance statistics was the model's aggressive appearance.

Critics branded it much too close to a racing car in the image it projected, especially coming from such a respectable company as BMW. Although tame by today's standards, the '73 debutant shocked commentators with its tacked-on trunk spoiler, side skirts linking the swollen wheel-arch extensions and their fat tires, and a deep race-style frontal air dam. Most notoriously of all, its colorful graphics included full-width "2002 turbo" script in mirrored writing on the frontal panel, just so that the driver ahead would get the message loud and clear.

In the ensuing outcry, the mirrored script was quietly deleted, but this was the least of the Turbo's troubles. It was

The Turbo dashboard was not very different, apart from the all-important boost gauge to the right of the console. The sudden rush of turbo power gave the model an unjustified reputation as being tricky to drive.

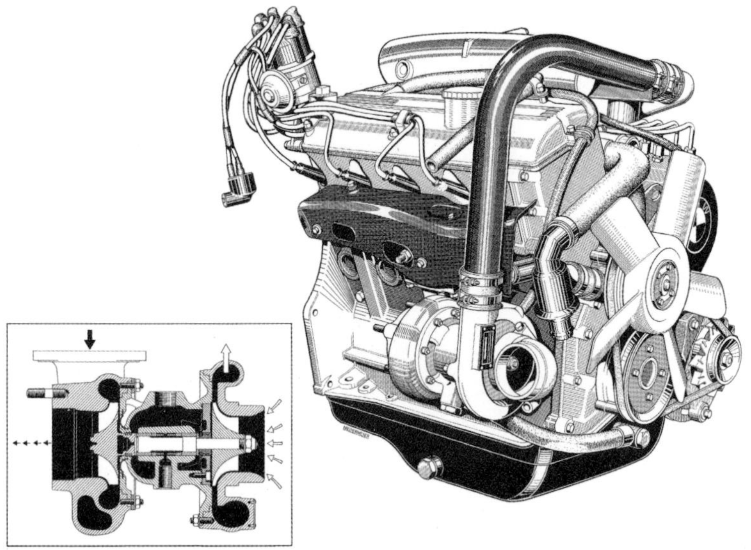

The turbo engine, with a cutaway of the turbocharger itself. Interest was immense as this was new technology for passenger cars in the early 1970s.

tricky to drive, especially on wet roads when the engine suddenly came on boost and the rear wheels lit up, but it was early days for turbo cars, and most drivers soon got used to treading carefully. The real blame lay elsewhere—and it was left to external circumstances to deliver the Turbo its fatal blow when hot politics in the Middle East led to a global oil crisis and a general panic about fuel prices and shortages. BMW, though braver than most at holding its nerve in a crisis, finally conceded that this was the wrong time to be promoting such a powerful model, and the 2002 Turbo was withdrawn after little more than a year on sale, with just 1,672 examples built.

GUARANTEED EXCLUSIVITY

There is nothing that car collectors prize more than an exotic halo model—one that was ahead of its time but is rare because it failed commercially through no fault of its own. The 2002 Turbo ticks all those boxes, and the few hundred surviving examples now command extravagant auction estimates.

The driver environment of the Turbo model was only slightly more luxurious than the cheaper models, and despite a price tag double that of a basic '02, the five-speed gearbox was an extra cost option.

A very period-looking publicity shot of a gleaming red '02, complete with optional spot lamps. The location shows how keen BMW was to emphasize the company's Bavarian roots.

BMW '02 SERIES: POWER TRAINS AND PERFORMANCE

Model	1502	1602	1602 ti	1802	2002	2002 ti	2002 tii	2002 Turbo
Engine Code	M10	M10	M10	M10	M10	M10	M10	M10
Configuration	Inline 4, SOHC 8V	Inline 4, SOHC 8V	Inline 4, SOHC 8V	Inline 4, SOHC 8V	Inline 4, SOHC 8V	Inline 4, SOHC 8V	Inline 4, SOHC 8V	Inline 4, SOHC 8V
Capacity (cc)	1,573	1,573	1,573	1,776	1,990	1,990	1,990	1,990
Fueling	Single carburetor	Single carburetor	Twin carburetor	Single carburetor	Single carburetor	Twin carburetor	Mechanical fuel injection	Fuel injection, turbocharger
Power @ rpm	75 @ 5,800	85 @ 5,700	105 @ 6,000	90 @ 5,250	100 @ 5,500	120 @ 5,500	130 @ 5,800	170 @ 5,800
Torque @ rpm (N m)	118 @ 3,700	123 @ 3,000	131 @ 4,500	143 @ 3,000	157 @ 3,500	178 @ 3,600	176 @ 4,500	240 @ 4,000
Transmissions	M4	M4	M4, M5	M4	M4, M5, 3AT	M4, M5	M4, M5	M4, M5
Max Speed (km/h)	157	162	175	167	173	185	190	211
0–100 km/h	14.5	13.5	11.0	12.0	12.0	10.2	10.0	8.0

The Turbo was a pioneer on many levels, not just because it brought forced induction to the general market. It marked the first tangible fruits of the embryonic BMW Motorsport operation, which was charged with not merely winning races—see the timeline on pages 7-8—but also exploiting those motorsport developments to enhance the road cars that people could actually buy.

The 2002 Turbo also highlighted the brilliance of the four-cylinder M10 engine, which had been designed by Alex von Falkenhausen's team at the beginning of the 1960s with the express intention of allowing for expansion and further development. Already, the 2.0-liter M10 was giving more than 240 horsepower in sixteen-valve nonturbo competition trim, enabling it to dominate Formula Two single-seater racing in Europe. Most famously of all, the same basic engine, sleeved down to 1.5 liters and highly turbocharged to suit the Formula One regulations, began winning F1 Grands Prix a decade later, taking BMW and Nelson Piquet to the World Championship crown in 1983. Relabeled M12/13 in its F1 application, the unit still enjoys the distinction of being the most powerful engine ever seen on a racetrack, offering "off the charts" outputs north of 1,400 horsepower in qualifying trim.

But it would be a mistake to see the 2002 Turbo as the most appropriate ambassador for the '02 generation. However technically fascinating it might have been, and however significant its development was for the nascent Motorsport and, later, M divisions, history would see the Turbo program as something of an aberration, an intriguing sideline to BMW's main business of building sporting cars to feed the growing market for quality models at realistic prices. And besides, the Turbo disappeared well before the rest of the '02 models were replaced by the 3-Series in June 1975.

Instead, it was the more mainstream 2002 ti and tii models that handed down the most valuable DNA to future generations of BMWs. Compact in stature but classy in their quality and composure, they proved that cars could be fabulous fun at the same time as being practical and untemperamental. It's a recipe that continues to inspire BMW to this day.

As the '02 line approached its tenth successful year in production, big changes were underway in BMW's Munich headquarters. Already, visionary CEO Eberhard von Kuenheim had been on board for a good three years, and the company was now on a much stronger financial footing thanks to the sustained strong sales performance of the Neue Klasse and the '02 models.

More importantly, von Kuenheim's long-term strategy was beginning to take shape. Centered on the three core model lines, which became the 3-, 5-, and 7-Series, it was a coordinated program to secure BMW's position in the most important market sectors. The 5-Series, which directly replaced the Neue Klasse models, had launched to some acclaim in 1972, despite the Arab oil embargo and international fuel shortages. The future 7-Series, still under development, was set to replace the big 2500–2800 sixes and take on Mercedes at its own game.

It was the momentum built up by the '02 model lines that provided the crucial impetus for all these programs, the most important of which was the 3-Series. And, as we will see in the chapters that follow, each of those many 3-Series generations in some way harks back to the energetic spirit of the '02 models that did so much to bring BMW to the attention of the whole world.

ABOVE: A deeply felt but very belated tribute to the 2002 tii came after the launch of the G80-generation M4 coupe in 2020 when New York fashion house Kith decided to add its high-luxury touches to both an M4 and its fifty-five-year-older spiritual inspiration, the 2002 tii. As the detailed photos show, the bespoke Kith 2002 was kitted out to luxury levels unheard of in the 1970s, including plush leather lining on most of the interior panels. The elegant alloy wheels echo the Alpina style of the period, and the tires are standard size.

M-MV 2263

SIMPLE BUT BRILLIANT: A STAR IS BORN

It's barely a few weeks into the 1970s and French-born car stylist Paul Bracq is arriving at BMW to take up his new post as chief designer. It's the first time since the late 1930s that the company has had such a role, and, as Bracq would soon find out, his appointment didn't come without in-house tensions.

In Bracq's briefcase is a highly impressive résumé: In his decade at Mercedes-Benz, he shaped the highly acclaimed 230 SL "Pagoda" sports car as well as the mid-'60s 250 SE, forerunner to the S-Class dynasty, and after leaving the prestigious German firm he had a hand in the design of the iconic French TGV high-speed train. The BMW company is on an upward trajectory, and it's easy to see why he is the right man for the job. But not everyone is happy with that.

In his in tray are three priority tasks. He needs to fine-tune a proposal from an Italian design house for a facelift to the '02 series, which is still selling six-figure volumes a year but looking its age. In a similar vein, design work on the upcoming E12 5-Series midsize sedan—the replacement for the brand's 1961 savior, the Neue Klasse—is nearly complete but needs some key finishing touches. But most important is Engineering Project E21, already underway for several years: the all-important replacement for the 2002 and its lesser brothers, which have been in production since 1966.

Onerous tasks, for sure, but BMW is in good shape. Its three main model lines are selling steadily, providing strong cash flow to fund the big investments coming up for the new Dingolfing factory and the E12 5-Series. And while other business leaders are noting the rising tensions in the Middle East with some foreboding, BMW's still-young CEO, Eberhard von Kuenheim, barely a year into his tenure, holds his nerve and stays on course with his bold plans to invest and expand. Central to that strategy are three core model lines, which will become the 3-, 5-, and 7-Series, and it is the 5 that will lead the charge.

THE ITALIAN CONNECTION

Wilhelm Hofmeister is the individual most frequently linked to BMW styling in the 1960s and 1970s, and indeed his name is industry shorthand for the signature kick-forward kink found at the base of the rear pillars of nearly every BMW model from the Neue Klasse onward. Yet in reality, Hofmeister's role was a technical rather than an artistic one: He was in charge of body engineering, frequently subcontracting aesthetic design work out to outside firms, many of them Italian. These external design studies would then be pitched against BMW's own internal proposals and a winning design selected. In 1966, Bertone had produced a proposal for the big sedan that would become the 1968 2500–2800; the Italian study had lost out to BMW's in-house design, but company officers were sufficiently impressed that they put Bertone on a retainer and noted the signature on many of the sketches—Marcello Gandini.

Bertone had better luck next time round, when its proposal for the more compact E12 5-Series was selected ahead of the internal offering, and it was this nearly finished design that greeted Bracq as he settled into his new role. Yet, as author and onetime BMW consultant Steve Saxty relates in his book *BMW's Hidden Gems*, the interior of the car was "a blank canvas," and Bracq was able to implement his radical new dashboard and driver interface concept—something that would distinguish all new BMW models from that moment on.

The balanced proportions of the first 3-Series are clear to see in profile, with the gentle wedge to the hood echoing the themes of the larger 5-Series, launched three years earlier.

The biggest change in style came at the rear, with the '02's oversize deck lid morphing into a much more harmonious sloping tail. Moving the license plate to below the bumper freed up space for broader rear lights and a decor panel.

Time warp: Veteran US journalist Karl Ludvigsen on the 3-Series debut drive in 1975. In his report for *Motor Trend*, Ludvigsen praised the new car's big step-up in quality, comfort, and refinement.

While all this was going on, BMW had circulated the engineering specifications for a third model, the E21 3-Series, around key design studios; among the responses was Bertone's, led by Gandini. As Saxty relates, having spoken with many of the individuals involved at the time, the Italian *carrozzerie* clearly saw Bracq's appointment as design director as something of a threat to their business with BMW. This prompted Bertone to build a fully finished show prototype similar to one of its proposals for the E12 5-Series and display it on the show circuit under the name Garmisch—which happened to be a town close to BMW's Munich headquarters.

Though the Garmisch (see chapter 10) was dimensioned as a possible replacement for the 2002, this was not the proposal eventually approved by the BMW board for the upcoming E21 3-Series. Instead, Bracq, who had been given a free hand for the new model's design, came up with his in-house proposal for the compact two-door, featuring a hatchback rear treatment similar to that of the 2002 Touring. Newly arrived sales chief Bob Lutz, who would go on to a glittering career with positions at almost every leading Western automaker, liked the design but vetoed the hatchback format, reputedly saying that German buyers prefer a big trunk—and Bracq was in no position to object. His design was masterful on many levels, and it was in this way that the unmistakable silhouette of the 3-Series, with its forward-slanted nose, long hood, and short trunk, was born. This profile immediately distinguished BMW's cars from their humbler competitors and established the template for quality sedan design that would power the company to the very top of the premium car market in the decades that followed.

BMW 3-SERIES E21: TIMELINE

1975	3-Series launched as 316, 318, and 320, all two-door sedans
1975	Fuel-injection 320i with 125 hp added
1977	6-cylinder 320 replaces 4-cylinder 320i
1977	Baur Cabriolet added
1978	6-cylinder 323i with 143 hp added
1981	Entry-level 315 added
1982	All models except 315 cease production as successor E30 version ramps up
1983	315 ceases production

BMW 3-SERIES E21: STYLES AND VERSIONS

Body Style	2-Door Sedan	2-Door Convertible
Length (mm)	4,355	4,355
Wheelbase (mm)	2,563	2,563
Curb Weight	1,040–1,180	1,090–1,110
Number Built	1,361,039	3,000

BMW 3-SERIES E21: BASICS AND ENGINEERING

Layout	Front engine, rear-wheel drive
Suspension F/R	MacPherson struts/semitrailing arms/coil springs
Steering	Rack and pinion; power assistance optional on 6-cylinder cars
Braking	Disc/drum

ENGINEERING, EASY—REGULATIONS, NOT SO EASY

With the external styling frozen and the major engineering elements carried over from the '02 series, E21 3-Series development was well advanced by the time its bigger brother, the 5-Series, launched in summer 1972, just as the Olympic Games were descending on Munich. But several complicating factors served to prolong the 3's gestation—most notably the tightening US air quality and safety legislation—and it wasn't until summer 1975 that the first versions hit the dealerships.

Responding to feedback from '02 owners, BMW made sure the new car had more room in the rear seats than its predecessors, leading to a longer wheelbase and larger trunk. The new instrument panel took the revolutionary and highly praised ergonomic design from the 5-Series and improved it still further, with a center console angled toward

(CONTINUED ON PAGE 48)

DESIGN UP CLOSE E21

This is how it all started. Paul Bracq's masterful design combined a forward-looking slight wedge profile with a neat two-door glasshouse between the short trunk and the longer clam-shell hood, leading to an early version of the famous BMW "shark nose," but with only mild plan shape. The vertical BMW kidney grilles just break the leading edge of the hood and sweep back into a subtle raised "power dome" that carries back on the hood before fading out ahead of the windshield.

The overall shape and proportions are fresh and modern, but they clearly reference the more upright and squarer 2002 series. Almost nothing changed visually on the E21 throughout its production lifespan, and the groundbreaking "cockpit" design interior stayed unaltered, too. It remained resolutely a two-door, and the only other body variant was the Baur Cabriolet.

The design is notable for its immaculate detailing and lack of unnecessary ornamentation, while visual differentiation across the range is kept to a minimum: Top-of-the-line versions such as the 320i and 323i featured four smaller headlights instead of two larger ones, and the 323i had a dual-line exhaust system with twin tailpipes.

The E21 3-Series design is of immense importance in the BMW story. More even than the first 5-Series, it established the template for compact quality sports sedans in general, and for BMW in particular it provided the rock-solid brand identity that would launch the firm onto a trajectory of unprecedented growth and see it become the world's leading producer of premium cars.

LEFT: The double kidney grille brand signature was subtly modernized for the E21 3-Series, becoming shallower, squarer, and with the two halves more closely fused. By gently breaking the front edge of the hood, the grille flows back into the raised hood ridge to draw attention to the engine underneath.

OPPOSITE TOP: Paul Bracq's masterful lines for the E21 3-Series were gently progressive for the time and set the template from which all subsequent generations have evolved.

OPPOSITE BOTTOM: The sloping rear deck line gave Bracq's 3-Series a much fresher look, with just the right balance between coupe-like sportiness and visual roominess. Immaculate detailing meant that almost no design changes were needed during the model's eight-year production run.

BMW Classic
M ST 3381

ABOVE: With large and beautifully clear instruments and controls, the wraparound driver environment of the 3-Series represented a major advance over its contemporaries—and became a strong selling point.

LEFT: The first 3-Series carried over the highly regarded mechanical elements from the '02 models. But with US smog regulations beginning to bite, North American export models were significantly down on power, torque, and sparkle compared with free-breathing European market cars.

(CONTINUED FROM PAGE 45)

the driver to create an intimate cockpit feel. This, too, would become a brand-defining feature passed on to all subsequent BMWs.

On the mechanical front, the vehicle structure was heavily stiffened to fulfill US impact rules, an ancillary benefit of which was the opportunity to dial down the suspension stiffness without losing handling precision; the switch to rack-and-pinion steering from worm and roller was an important move in that direction, too. Inevitably, however, the stronger structure, the extra equipment, and the larger footprint told their story at the weighbridge, sending the needle some 50 kilograms further round the dial. BMW quickly pointed out that the launch engines—a 1.6-liter, which made 90 horsepower in the 316, and a 2.0-liter, which made 109 horsepower in the 320 or 125 horsepower in the fuel-injected 320i—were more powerful and more flexible than their '02 predecessors.

SUMMER '75: HIGH EXPECTATIONS—AND MIXED REACTIONS

Three years beforehand, the 5-Series had been launched to considerable acclaim for its fresh style, groundbreaking interior ergonomics, comfort, and refinement. The only dissenting voices came from the hard-driving enthusiast community, some members of which declared that the 5 had gone soft in pursuit of a broader client base, supposedly betraying BMW's sporting values in the process.

Those complaints were, in any case, soon settled with the arrival of the quicker six-cylinder 525. Overall, the 5-Series had set the bar pretty high, so the expectations surrounding the 3-Series were, if anything, higher still. And, broadly speaking, the reaction to the new car showed it fulfilled those expectations. The response of US commentators was particularly important, given the significance of North America in BMW's export strategy and the fact that American buyers had so revered the 2002 family.

Revealing, here, are the words of veteran US auto journalist Karl Ludvigsen, later a prominent industry analyst, who in July 1975 was sent to Munich by *Motor Trend* magazine to sample the new model. Here are some of the more perceptive passages from his report:

> [The new models] supplant a line of small sedans that have earned a worldwide reputation for their handiness, convenience, toughness and fine performance. But is this new car good enough to make their owners want to trade them in on it? BMW hopes so, because the 3-Series is essentially the '02 series car changed only as needed to make it better. . . .
>
> Finished entirely in sullen flat black, the dash and instrument cluster give an impression of highly self-conscious efficiency. . . .

On a recent visit to Munich, I drove both the 320i, with Bosch K-Jetronic fuel injection, and the 320A, equipped with a smooth-shifting new ZF automatic transmission (model 3 HP 22) behind its Solex-carbureted engine, which delivers 80 horsepower against the 320i's 92 horsepower. Both cars were noticeably quieter, with less engine, wind, and road noise, than the '02-series cars.

"Will these features and performance keep BMW fans happy with the 3-Series?" asked Ludvigsen by way of conclusion. "I think they will."

Very noticeable in Ludvigsen's report are the horsepower figures he gives for the US versions. From the perspective of today's harmonized engineering and emissions standards, it is easy to forget how cruelly strangled these cars were by the manufacturers' early de-smogging solutions. The US-spec 320 sampled by Ludvigsen offered 80 horsepower, compared with the unrestricted European car's 109. For the sportier, fuel-injected 320i, the gulf seemed even wider: 92 horsepower versus 125.

(CONTINUED ON PAGE 52)

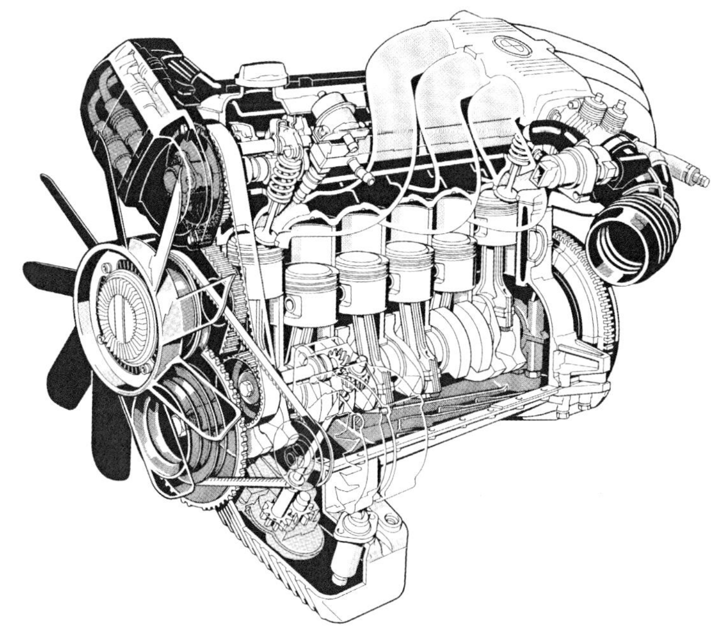

The super-smooth six-cylinder M20 engine marked a key step in the 3-Series' advance in power and status. The carburetor-fed 320 six was actually less torquey than the four it replaced, but the 143 hp fuel-injected 323i was a real cracker and gave new meaning to the term "sports sedan."

Confirming the new 3-Series' role as a quality sports sedan it boasted many of the tucked-away details of BMW's bigger models, such as the built-in toolkit in the trunk lid and the plug-in rechargeable flashlight in the glove box.

MOTORSPORT AND THE E21

Were it not for the remarkable record of the mighty 3.0 CSL "Batmobile" coupes in the epic European Touring Car Championship battles throughout the 1970s, the equally heroic antics of the smaller BMWs—principally the 2002 ti and the later 320—might be more clearly remembered. The spectacular CSLs swept everything before them, driven by aces such as Dieter Quester, Manfred Winkelhock, Hans Stuck, and Ronnie Peterson, but they were in the premier class with big six-cylinder engines.

Lower down the pecking order, BMW 1602s and 2002 tis had been competing with considerable success in classes I and II, respectively, and just before turbocharging was outlawed in 1970, the 2002 TIK had taken the overall Touring Car title.

The 2002 in its various incarnations kept the BMW flag flying in classes I and II while its bigger CSL brother built up to a dominant position from 1973 onward, taking six out of seven overall championships in the decade. But the smaller BMWs, entered in large numbers by independent teams, were never far behind. In 1975, the Heidegger Racing team even entered a 2002 ti in the Le Mans 24-hour classic; it qualified 55th but lasted the course, finishing a heroic 27th overall.

It was some time after the road-going E21 3-Series launched in 1975 that racing versions appeared on the tracks. While BMW's official focus was on factory-supported teams such as Alpina and Schnitzer campaigning in the big-capacity class, by 1977, BMW's

Race-tuned versions of the 320i first took to the circuits well after the road car's debut in 1975, forming the mainstay of the BMW Junior Team. The 2.0-liter, sixteen-valve engines gave in excess of 300 hp and also became the dominant engine in Formula 2.

own Junior Team was fielding the new Group 5 320 racer in the 2.0-liter class, with pilots including Marc Surer and Eddie Cheever. The lightweight cars, whose sixteen-valve, 300-horsepower engines were already the power unit of choice in Formula Two single-seater racing, won on their first outing at Zolder. From then on, they were the cars to beat, taking the manufacturers' world championship the following year.

One of the 320's most spectacular results was ninth place overall in the 1977 24 Hours of Le Mans with Hervé Poulain at the wheel. A turbocharged version of the 320i was fielded by McLaren North America in the late 1970s with a star driver lineup, taking seven wins, but it was only when Group 5 regulations loosened up in Europe that things got really wild. BMW Motorsport sold Group 5 kits to customers wishing to race their 320s, and this was the era of huge rear wings, vast squared-off wheel-arch extensions, and ground-hugging cowcatchers up front. The silhouettes of the racers seemed to bear little relation to the road cars they were derived from, least of all the Ford Capris and BMW 320s. The engines, now just 1.4 liters but highly turbocharged to give more than 450 horsepower, made the legendary orange Jägermeister 320s difficult to drive—but it was all part of a grand BMW plan to gain on-track experience with extreme turbocharging in preparation for an assault on the "Königsklasse," Formula One, in the seasons to come. And that, as has been amply documented, resulted in not only the most powerful engine ever seen in Formula One but also the 1983 World Championship crown going to Brabham-BMW's Nelson Piquet.

Marc Surer was one of the first recruits to the BMW Junior Team in 1977. He, along with teammates Eddie Cheever and Manfred Winkelhock, enjoyed spectacular success and later graduated to higher formulae, including F1.

Jochen Neerpasch (left) and Markus Flasch in front of the Group-5 320i. As the founding father of BMW motorsport, Neerpasch launched the factory Junior Team in 1977. Today, the juniors' starter car is the M4 GT4.

The car that everybody coveted in the 1970s: The 323i was a magical distillation of the best BMW values—a fabulously smooth and energetic six-cylinder engine, razor-sharp handling, and a quality cachet that no other compact car could match.

The interior of all first-generation 3-Series cars was simple but beautifully executed, setting standards for the compact sector. Innovations included the check control and the service interval indicator. No four-door option was ever offered.

(CONTINUED FROM PAGE 49)

There were no such problems for Britain's Clive Richardson, testing a 320 for the January 1976 edition of *Motor Sport* magazine:

> That smooth, four-cylinder engine revels in town work as much as it enjoys high speed, pulling casually from 1,000 r.p.m. in third gear. Speed up a little and the 320 shows remarkable agility in traffic, the steering responding instantly to quick changes of course, shrinking the car's true size. The squarish front-end shape and good visibility make it an easy car to "place."

Richardson went on to lavishly praise the new car's massively improved handling balance, though he was less happy with the low overall gearing and the fuel economy. Overall, he concluded, "there is no mistaking the 320 for a BMW: that feel of tautness, the aura of quality, a train-like sensation of dependability and attention to detail. Added to that, this new model is comfortable, relaxing and quite extraordinarily easy and enjoyable to drive. BMW Concessionaires GB, in the past so heavily criticized for overpricing their range, could almost be accused of charity, pricing the 320 at £3,349."

Richardson's comment on the BMW's pricing went against the popular perception of the company's marketing strategy at the time. Indeed, at a press conference I attended some years later, Anton Hille, chief executive of the UK importers, swaggered openly, "Our cars aren't expensive—they're *very* expensive."

Even without the badge, this clearly identifies as the top-line 323i with its twin exhausts, sometimes referred to as wheelbarrow handles. The 323i replaced the Golf GTI as the company car most craved by ambitious city high fliers.

If there was criticism of the new 3-Series in any quarter, it was that the car's overall style was perhaps too cautious, too derivative of the bigger 5-Series. But with half a century's hindsight, those reservations seem misplaced. Bracq's 3-Series shape might have been neat and unthreatening, but its subtle wedge profile and its harmonious linking of front and rear were ahead of their time—and indeed, these attributes helped establish a typology that would inspire many future generations.

SKYROCKETING SALES

Any prelaunch reservations about the new car's move into classier premium territory were dispelled within the first few months on sale: Transactions immediately began running at almost double the rate of the outgoing '02, and by the end of its first six months, more than 44,000 had found eager buyers. Sales nearly tripled the following year, broke the 200,000 barrier in 1980, and hit a peak of 226,960 in 1981. These were figures unprecedented in BMW's history.

All in all, by the time the E21 handed the 3-Series baton to its E30 successor in November 1982, it had sold a remarkable 1,364,039 units, making it not only BMW's fastest-ever seller but also its first model to top a million units—and by some margin, too.

As befits a right-first-time design, there were almost no changes to the 3-Series' appearance or mechanical makeup over the course of its nearly eight-year lifespan—except for one very significant strategic move. Sensing that the 3 had the qualities to climb still higher in the market, BMW had begun developing a new range of small-block six-cylinder engines. The M20 was an all-aluminum overhead-camshaft twelve-valve design, with initial capacities of 2.0 and 2.3 liters. Its first application came as a carburetor-fed 320, giving 122 horsepower—fractionally less than the four-cylinder model that it gradually supplanted. And while critics praised its sweetness and smoothness, almost all said that its low-down flexibility fell short of what they were used to with the four. The villain here was the 80- to 90-kilogram hike in weight brought on by the larger unit. But customers weren't deterred: The 320 six quickly became a bestseller, outselling all but the gateway 316 over the lifespan of the E21 series.

Basking in the success of the sophisticated six-cylinder and the classy cachet that it projected, BMW nevertheless had another card up its sleeve—and it proved to be a game changer. Making its debut in the first weeks of 1978, the 323i initially seemed to be just another engine variant, but it quickly proved to be one of those rare models that is more than the sum of its parts. Everything gelled together

(CONTINUED ON PAGE 56)

COLLECTABILITY

This first-ever 3-Series is certainly a highly collectible model, but even the youngest survivors are now more than forty years old, and as with the '02 series, rust is still the biggest enemy. Corrosion-free examples are hard to find, and restored cars are pricey—but all models, even the humble 316, provide an engaging drive and a reasonably dependable everyday companion.

Standout model: 323i. This car, with its cracking 143-horsepower straight six, set the template for sophisticated sports sedans, but many examples have been modded or abused.

Hidden gem: 320i. Lovely four-cylinder engine, fun chassis, but overlooked in favor of the fashionable sixes.

Wild card: Original Baur convertible, despite its ungainly roll bar. Unloved in period, rare and rust-prone, but less likely to have been abused in its prime.

The hero car they all want: The top 323i is the fastest and most fun, but beware tuned and abused examples, and check carefully for rust or poor restoration work.

Hidden gem: Often overlooked in the frenzy for the six-cylinder 323i, the earlier four-cylinder 320i is a sweet, fuel-injected delight—if you can find one that's been kept unmolested for its forty-five-plus-year life. The big bumpers betray this as a de-smogged US-market version, with reduced power and torque.

ABOVE: Something of a wild card choice is the Baur convertible. Not pretty and not as responsive as the lighter sedans, but less likely to have been thrashed or modified. As always, check for corrosion—they all suffer.

LEFT: The driver-focused cockpit of the first 3-series represented a massive advance just for BMW but for automotive design as a whole. For the first time, vital information was clearly and attractively presented, with impeccable high-grade materials and finish. It was a big hit with buyers and set a lasting trend across the industry.

As basic as it gets: Twin, larger headlights reveal this as a four-cylinder, noninjection car—but even as an entry-level model the 316 was well built and fun to drive, though equipment levels were sparse.

(CONTINUED FROM PAGE 53)

magically, providing a vivid 143 horsepower from its silky-smooth, free-spinning six and Bosch fuel injection giving razor-sharp throttle response right up to its six-thousand-plus red line. This, allied to its slick-shifting five-speed gearbox and eager steering response, made it fabulous fun to drive. What's more, though it was more than capable of tangling with the hottest of hot hatches, its classily sober exterior made it the weapon of choice for ambitious, image-conscious executives climbing the corporate ladder. At last, here was a worthy successor to the fêted 2002 tii.

At a stroke, BMW had not only raised its own game but also—and more significantly—invented a wholly new class of vehicle: the premium small car. No longer did compact cars have to be dull, flimsy, and unappealing. Now, suddenly, they could be just as classy, luxurious, and refined as big luxury models, and they offered the added bonus of being cheaper to buy and run than their full-size counterparts as well as much more fun to drive. Which was fast becoming BMW's official mantra.

REASONS TO SUCCEED

So, why did the first 3-Series turn out to be such a hit? In truth, the model's runaway success in its first few years surprised even BMW's meticulous planners, who had been expecting a decent improvement on the '02 series' sales performance, but not a doubling of it. It was clearly the right car for the moment, offering buyers with modest budgets a taste of the comfort, the kudos, and the fun that had previously been the preserve of the wealthy customer. On a social level, it offered a handy leg up into a more sophisticated set of classy people in the know.

Another success factor may have been BMW's singular focus on just the single two-door configuration, thus guaranteeing a strong and consistent brand message and widespread recognition. There had been pressure for other versions, but the only one to make production—albeit at the hands of an external contract manufacturer—was the Baur Cabriolet. This was not a very successful conversion, with a clunky roll bar, an awkward two-part top, and a significant weight penalty that blunted its performance. Only a few thousand were built.

BMW 3-SERIES E21: POWER TRAINS AND PERFORMANCE, NON-US MODELS

Model	315	316	318	320	320i	320/6	323i
Engine Code	M10	M10	M10	M10	M10	M20	M20
Configuration	Inline 4, SOHC 8V	Inline 4, SOHC 8V	Inline 4, SOHC 8V	Inline 4, SOHC 8V	Inline 4, SOHC 8V	Inline 6, SOHC 12V	Inline 6, SOHC 12V
Capacity (cc)	1,573	1,573	1,754	1,977	1,977	1,990	2,315
Induction	Single carb	Single carb	Single carb	Single carb	Bosch mechanical injection	Single carb	Bosch mechanical injection
Power @ rpm	75 @ 6,000	90 @ 6,000	98 @ 5,800	109 @ 5,800	125 @ 5,700	122 @ 6,000	143 @ 6,000
Torque @ rpm (N m)	110 @ 3,000	123 @ 3,000	142 @ 4,000	157 @ 3,700	172 @ 4,350	160 @ 4,000	190 @ 4,500
Transmissions	M4, AT3	M4, AT3	M4, AT3	M4, AT3	M4	M4, AT3	M4, M5, AT3
Max Speed (km/h)	160	161	168	173	182	183	192
0–100 km/h	14.0	13.0	12.0	11.5	10.0	10.5	9.5
Comments	Entry-level runout model						

Many bemoaned BMW's failure to produce a four-door version, or a station wagon or coupe, all of which would appear in later generations. But the real losers in the first-generation 3-Series stakes were North American buyers—which was particularly galling because they were the ones whose enthusiasm for the 2002 family had put BMW on the map in the US. Not only did they suffer the ignominy of bulky impact bumpers front and rear, but they had to put up with sometimes temperamental and underpowered de-smogged engines. To cap it all off, they were denied the delicious six-cylinder engines, instead being short-changed with a fuel-injected 1.8-liter four, still badged as 320i to pacify those who valued the prestige of a higher-powered label.

But BMW had not been idle, and much work had been going on behind the scenes. All these deficiencies, and more, would be remedied with the arrival of the second-generation model, designated E30, as 1982 rolled into 1983. Yet the original car would enjoy one last curtain call, even after its successor had gone on sale: Simplified 315-badged versions stayed in production throughout 1983 to provide a lower-cost entry point into BMW ownership as the new, and more expensive, car found its feet in the market.

All in all, to this first-generation 3-Series goes the honor not only of being supremely successful, beyond BMW's wildest dreams, but also (more significantly in terms of the dynamics of the global car market) of inventing the market for premium compact cars. This historic shift would reshape the international car market forever—and it put BMW firmly onto the path of prosperity and rapid growth that would see it become a global player and a top dog in the market for premium cars.

Right from its launch in 1975 the baseline 316 was by far the strongest seller in the 3-Series lineup. In the E21's run-out phase BMW offered a detuned 315 version to tempt eco-minded buyers: it sold more than one hundred thousand units over three years.

BMW
M·MX 4775

SMARTER AND FASTER: BMW SEIZES THE ADVANTAGE

The second-generation 3-Series was the first to offer four doors and, later, a Touring station wagon. Though giving a similar first impression to the outgoing model, it was in fact larger and different in every respect.

TOP AND ABOVE: Understated in every respect, the second-generation E30 3-Series was initially criticized for being too cautious, but went on to become an even bigger success than its predecessor. It was the first generation to offer multiple body styles and drivetrain types.

By the end of the 1970s, BMW was getting serious. Its big bet, the E21 3-Series, was selling almost two hundred thousand units a year, more than double what the whole company had been producing just a decade back. The big 7-Series had taken the fight to archrival Mercedes-Benz, and a fresh second-generation 5-Series was at an advanced stage of development. And, secretly, top executives knew that their motorsport division was gearing up for an assault on the most prestigious race series of all—the Formula One World Championship—using the production-derived, turbocharged four-cylinder engine that was beginning to notch up the wins for the 320i racer in the European Touring Car Championship (ETCC).

But CEO Eberhard von Kuenheim, never prone to complacency, knew full well that his counterparts at Mercedes must be stung by the success of the 5-Series and fearful for the impact of the 7-Series on Mercedes's highly profitable S-Class halo model. BMW knew, too, that the spectacular success of the 3-Series had prompted the rival firm to begin development of a smaller model aimed directly at the 3.

So the mission of the second-generation 3-Series was much more than simply to consolidate the success of the first. The new car also needed to broaden the 3's reach and strengthen its capabilities in order to counter the likely assault from the compact Mercedes. It also needed to be faster, more comfortable, safer, cleaner, and capable of accepting a much wider variety of engines and transmissions.

CAUTIOUS EVOLUTION

That sort of task list will be familiar to anyone who has been involved in car design and engineering: seemingly out-of-reach targets, conflicting objectives, and, invariably, limited resources. But BMW's by-now highly respected engineering talent managed to tick all those boxes, and more, with the new design. And, what's more, it would be an even bigger success than its forebear.

Yet when the wraps came off in autumn 1982 and the new E30 was revealed as a cautious, evolutionary shape, there was a mixture of surprise and disappointment. Surprise because the second-generation 5-Series, launched the year before, had been roundly lambasted for its near-indistinguishable styling revamp—although that hadn't deterred customers one bit. Disappointment because BMW's new design director, Claus Luthe, had been the man responsible for the profoundly futuristic NSU Ro80 streamliner a decade before. Armchair commentators had expected more. But as is so often the case with BMW's designs, closer examination of the E30's execution revealed the depth of its expertise.

BMW 3-SERIES E30: TIMELINE

1982	Second-generation 3-Series launches as two-door 316, 318i, 320i, and 323i (139 hp)
1983	Four-door body style and high-efficiency 325e added
1985	325i with 171 hp replaces 323i; four-wheel-drive 325iX announced; 324d and 324td diesels added
1986	Factory-built 320i Cabriolet replaces limited-numbers Baur conversion model
1986	First M3 launched, with 195-hp 16-valve 4-cylinder engine derived from M5 unit
1987	Mild exterior facelift; 318i and 324td gain motor electronics
1987	Touring station wagon announced, with production to begin in January 1988
1988	M3 Cabriolet added; M3 Evolution has 215 hp
1989	Two-door 318iS with 16-valve engine and 136 hp added
1990	M3 Sport Evolution raises power to 238 hp
1990	New-generation E36 model announced; E30 sedan models cease production
1993	Cabrio production ceases
1994	Touring production ends

BMW 3-SERIES E30: BASICS AND ENGINEERING

Layout	Front engine, rear-wheel drive, or four-wheel drive
Suspension F/R	MacPherson struts/semitrailing arms/minibloc coil springs
Steering	Rack and pinion, power assisted (except 316 and 318)
Braking	Disc/drum, servo assisted (316 and 318); disc, ABS optional

BMW 3-SERIES E30: STYLES AND VERSIONS

Body Style	2-Door Sedan	4-Door Sedan	4-Door Touring	2-Door Cabriolet	M3	Grand Total, All Models
Length (mm)	4,325	4,325	4,325	4,325	4,360	
Wheelbase (mm)	2,570	2,570	2,570	2,570	2,562	
Curb Weight	1,020–1,190	1,045–1,270 (325iX, 324td)	1,120–1,290	1,150	1,285	
Number Built	1,058,562	908,789	103,704	143,425	17,970	2,339,248

The new model's initial engine lineup was mostly carried over from the previous model, with updates to reflect tightening emissions regulations. Later, a wide variety of new power trains would join the range.

In many markets BMW, along with other German manufacturers, faced criticism that their cars were poorly equipped for their price. These standard steel 3-Series wheels look penny-pinching to today's eyes but were well accepted at the time.

A novel initiative from BMW on the 3- and 5-Series from 1983 was the "eta" engine, a high-efficiency, low-revving six. Super-smooth and quiet, and though it didn't fit BMW's sporty image it did attract a good following in North America.

The BMW grille continued to evolve, becoming shallower and broader for this second 3-Series. The twin kidneys are now slightly separated again, and the "power dome" running down the hood is now broader.

BMW's first forays into diesel power did their best to be smooth and powerful, but it took the turbocharged 324td to deliver anything like the sporty feel BMW buyers were expecting.

Luthe's sharpened-up evolution of Bracq's familiar 3-Series design and proportions successfully freshened the shape for the new decade, while the wider tracks and bigger wheel and tire combinations gave it a more firmly planted stance on the road. The lower beltline allowed more emphasis on the taller glasshouse, giving a less coupe-like profile that hinted at greater maturity and more space inside for passengers. And inside, the driver's environment was further upgraded from that of the highly praised original (again the work of Bracq) with a suite of advanced electronics that placed BMW in the vanguard of technical innovation at the time. Cascading down from the 7- and 5-Series before it, the new 3-Series featured a check control to monitor vital functions, a trip computer, an indicator that calculated when the next service was due, and an energy meter that gave a real-time readout of fuel consumption. All were new to the compact sector.

On the mechanical front, there were important changes to the rear suspension to tame the predecessor E21's sometimes wild tendency to suddenly lose grip at the rear in wet conditions or when driven to its limits. The lineup of engines carried over from the outgoing model now all had gasoline fuel injection apart from the entry-level 316, but in typical BMW fashion that most basic model would be the first to gain the all-new M40 engine and injection in 1986 to become the overall bestseller.

To start with, only a two-door body was offered, but neither that nor the conservative presentation seemed to deter buyers: The 318i took over one-third of the line's 220,000 sales in its first year and quickly established itself as the top seller. Mercedes's W201 190 appeared as a four-door in late 1982, but BMW had anticipated its big rival's next move, and a four-door variant of the 3-Series was already on the stocks and ready to launch for the 1984 model year. The effect was immediate: a further rocket boost to sales, taking the annual total to more than 320,000, and a firm foothold in the business and family car sectors, where serious-looking sedans were still regarded more favorably than coupe-like two-doors.

DESIGN UP CLOSE E30

As the 1970s rolled into the '80s, the original E21 3-Series was beginning to look out of step with its peers. It came across as tall and narrow-tracked, with a now-unfashionably aggressive shark nose, and the only alternative to the coupe-like two-door sedan was an ungainly convertible conversion. It had been spectacularly successful, especially among fun-focused drivers in the know, but its staid lines no longer advertised its prowess strongly enough to appeal to the broader market.

Great were the expectations accompanying the release, not a moment too soon, of Claus Luthe's E30 replacement model in 1982. Luthe had, after all, been responsible for the groundbreaking NSU Ro80 a decade earlier, and the cautiousness of his 1981 refresh of the 5-Series had been put down to budgetary restrictions. The big-numbers new 3-Series would be much bolder, it was suggested. Yet when the covers came off late in 1982, it, too, appeared to be more of the same.

Closer study of the E30 shape reveals the skill of designer Claus Luthe in modernizing the concept while retaining the feel of the original. The front view shows the fresher wedge profile, the shallower grille, and the wider power ridge on the hood drawing the eye forward from the screen to the grille.

But once more, its later success served to prove the value of careful evolution in building and enhancing brand identity. At first glance quite similar to the old, the key frontal graphics of the new car sharpened up the overall look, integrating the signature quad headlamps and the now flatter and squarer kidney grilles fully into the horizontal black grille backdrop; no longer did the kidneys break through the top and bottom of the backing, and the turn signals migrated below the bumper to allow the headlamps to be set wider for a feeling of greater width.

Most of all, this new design spelled solidity and security at every level. With wider wheels and tracks, it had a much broader and more stable stance on the road; the lower waistline made it look less top-heavy; and the neat, restrained detailing and excellent fit of the tires within the wheel arches gave a confidence-inspiring sense of engineering integrity to provide a powerful boost to the 3-Series' brand identity.

WIDENING THE 3'S APPEAL

Central to the commercial success of this second-generation 3-Series was the way that it consciously moved beyond a narrow focus on sportiness and appealed to a wider cross-section of buyers who could be drawn into the BMW fold on the strength of the brand's smart style, rock-solid build quality, and technical innovation.

First up was a very classy full Cabriolet, built in-house and replacing the gawky conversion by coachbuilders Baur. This immediately established the 3-Series among a wealthier, image-conscious clientele, but it was an agonizingly long wait until 1988 for the Touring station wagon. Reviving the name from the ill-starred '02 series derivative that had flopped in the marketplace, the 3-Series Touring wasn't the biggest or the most practical on the market, but it again secured a fresh tranche of buyers and created a new subsegment in the compact class—this time for "lifestyle" station wagons that combine quality engineering with sophisticated style to appeal to well-heeled and adventurous families. And, for the first time in the brand's history, BMW could draw in owners who needed to accommodate dogs or paraphernalia for outdoor pursuits.

All the while, BMW engineers had been busy in the background upgrading the models' technical credentials. The top engine jumped from 2.3 to 2.5 liters in the new 325i, with 171 horsepower and sparkling performance—and, for the first time, diesels entered the lineup in the shape of the 2.4-liter, six-cylinder 324d and turbocharged 324td. With just 86 and 115 horsepower, respectively, these seem feeble units by today's sophisticated standards. But in their time, they were revelatory in their smoothness, especially when compared with the tractor-like roughness of Mercedes's four-cylinder diesels; for a while, the same turbo engine in the bigger 5-Series held the title of the world's fastest diesel. Though the diesel models sold slowly, they were an important toe in the water for BMW and prepared the ground for the breakthrough direct-injection diesels that would offer more power than gasoline engines allied with even lower fuel consumption.

Lavish praise was heaped on the E30's interior, which took the predecessor's driver-focused format but made it even neater and more user friendly. Notable advances included the fuel consumption meter and, on top models like this M3, the trip computer.

FOUR-WHEEL DRIVE AND OTHER VARIATIONS

Something of a footnote in the E30 saga were the 325iX models, which introduced all-wheel drive to the 3-Series. The X models were distinguished externally by their swollen wheel arches and increased ride height, while underneath was a somewhat overengineered driveline with complex arrays of differentials and clutches to control the distribution of power to each wheel. Disappointingly to sports-minded buyers, this did not result in compelling rally-style excitement along the lines of an Audi Quattro or Lancia Delta Integrale, but the model did find favor in the more mountainous regions in Europe and snowbelt states in the US.

The 325iX was a relatively late arrival stateside, where the 3-Series lineup for the North American market had been in constant flux as the importers played with engines, specifications, and designations. One standout was the 325e, which employed a special version of the 2.7-liter six tuned for maximum efficiency thanks to a combination of a high 11:1 compression ratio and low revs. This had flopped in Europe, where buyers expected an eager, sporty edge, but its lazy, relaxed demeanor proved much better suited to the gentler US driving style.

RACING SENSATION: THE M3

If there is a single model that best symbolizes the second-generation 3-Series, it must be the M3. First unleashed in 1986, it was unlike any other 3-Series before—or since. Even though it was double the price of the top 325i and sold fewer than eighteen thousand units—a paltry 0.7 percent of the E30's total production of more than 2.3 million, itself a new record for a BMW—the ur-M3 is the standout model of its generation. This meticulously crafted car, which was far more different under its skin than its reassuringly familiar appearance might have suggested, was sensational to drive on both road and track and can now be recognized as the true father of the whole M bloodline and the M culture that persists to this day. Even now, enthusiasts still hanker after the raw and responsive thrills that the E30 M3 delivered so vividly almost forty years ago. And with so few built, it is highly prized among collectors and a surefire investment for those with an eye on a fast-appreciating asset. For more details, see the Collectability sidebar on page 72.

Jewel in the BMW crown: Most prized of all E30 M3s is this, the ultimate Sport Evolution model. Only six hundred were made over a three-month period in 1990. Today, examples change hands for six-figure-dollar sums.

The red plug leads identify this as the Sport Evolution model, the ultimate expression of the first-generation M3. BMW engineers increased capacity to 2.5 liters, revs to over 7,000 and power to 238 hp.

318iS: A LATE FOOTNOTE

With the M3 priced in the stratosphere and each successive version ramping up in power and exclusivity, other manufacturers were beginning to exploit the new niche for compact and affordable high-performance derivatives—such as Volkswagen with the Golf GTI 16V. BMW needed something to offer these sports-minded customers on less extravagant budgets and launched the 318iS in September 1989.

As was the norm with BMW, the new version looked little different from the regular model outwardly. But under the hood, there was a new engine that turned the car into a real—and often unrecognized—gem. A new twin overhead-cam, sixteen-valve cylinder head transformed the sweet and refined 1.8-liter into an eager athlete with an energetic 136 horsepower and undiminished sophistication. Paired with the slick five-speed transmission, it made for a very engaging driving experience, boasting better agility than the six-cylinder cars with more weight

The second-generation 3-Series hosted many drivetrain innovations, including all-wheel drive on the 325iX, distinguishable by its raised ride height and flared wheel arches. But despite the 4WD's complexity, the 325iX lacked sparkle when driven.

up front. The big mystery is why BMW seemed to produce the 318iS only as an afterthought, barely a year before E30 sedan production switched over to the new-look E36 third-generation model.

The 318iS should not be confused with the 320iS, which is another curiosity in the E30 story. This was essentially a variant of the M3, but with its engine size reduced to below 2.0 liters for the Italian and Portuguese markets with their punitive capacity-based taxation regimes. Incredibly, BMW M engineers were able to wring 192 horsepower out of the 1,990cc, sixteen-valve unit at a wild 6,900 rpm. It proved a worthwhile exercise, as almost four thousand cars found buyers.

Another oddity worthy of mention is the 333i. This was a purely South African derivative, built only in small numbers at the BMW satellite plant in Rosslyn, near Pretoria. In collaboration with BMW tuner Alpina, the South Africans slotted the 7-Series' "big-block" 3.2-liter six into the 3-Series shell, and all examples were right-hand drive.

A SMOOTH HANDOFF TO THE NEXT GENERATION

The E30 was on an absolute high in late 1990 when the principal sedan versions began to make way for their E36 successors. Production was down only slightly from its daily peak of more than 1,200 two years previously, and with overseas complete knockdown assembly numbers taken into account, almost 2.1 million had found buyers.

The Cabrio and Touring models would continue to be built for an additional three and four years, respectively, bringing the overall total for this exceptionally long-lived model to more than 2.3 million. The company had once again set new records and vindicated its strategy of diversifying into different body styles to expand its customer reach.

So, in conclusion, can the E30 generation be counted as an unqualified success? Very nearly. In true BMW fashion, it was expertly engineered and right for its time, despite early criticism of its cautious style. The M3 had spectacularly

(CONTINUED ON PAGE 71)

MOTORSPORT AND THE M3

The story of the E30-generation 3-Series in motorsport is a simple one—it's that of the M3. With its pure race-car blood, it set such a high bar that subsequent M-cars struggled to match its extraordinary sense of driver engagement.

There had been M-cars before the M3, of course, with the big E24 M635CSi and E28 M5 sedan both concealing staggering performance and remarkable refinement beneath sober, unostentatious, business-friendly bodywork. Both enjoyed a small but select following, but overall, they were low in profile and sought to disguise rather than advertise their prowess. The M3 was the polar opposite. It was a rebel, pure and simple.

For a start, it was unashamedly a racing car, developed by BMW Motorsport for the racetrack and only mildly tamed to make it suitable for everyday roads and everyday customers. Its singular focus was winning race series such as the German DTM championship, the prestigious European Touring Car Championship, and other touring car races around the world. Its role as a road car was a secondary one, and it only came about because the race regulations in the 1980s demanded that five thousand identical examples be built and sold to customers before the car could be homologated for racing.

The engine, a 2.3-liter, sixteen-valve dual-overhead-cam four, was developed by M engineers using the familiar M10 block but topped by a four-valve head derived from that of the M1's and M5's 3.5-liter six. The bodywork was flared outward, Quattro-style, to accommodate the wider wheels and race suspension, and a reangled backlight and reshaped trunk lid gave a higher deck line to improve airflow and allow the freestanding full-width rear wing to generate downforce.

As a road car, the M3 is easy enough to drive gently, once you're used to the tricky "dogleg" close-ratio gearbox, which places first back and to the left. The ride is hard, the car following every contour of the road, and the steering is lightning quick; the power delivery is urgent and instantaneous, demanding a

Exhilarating, engaging, addictive—all these and more apply to the legendary M3 in all its many forms. Originally developed as a race car, it was tamed just slightly for road use and quickly made faster again. This, the Sport Evolution, is the ultimate expression of one of BMW's most brilliant brand heroes.

delicate foot on the gas pedal. Out on the open road, though, this slight nervousness morphs into something much more engaging—a remarkable agility and responsiveness, akin to that of a track car but more benign. With some 200 horsepower readily available, and with little more than 1,200 kilograms to motivate, this car is rapid and responsive, even by today's turbo standards.

BMW began its factory support for several teams running M3s in 1987 and sold race kits to privateer drivers, too. The Motorsport-developed engine initially gave some 300 horsepower at 8,200 rpm, and the wins came thick and fast, with the Schnitzer team taking the European Touring Car Championship as well as many national series. An M3 even won the Tour de Corse leg of the World Rally Championship.

Successive years saw further performance-improving parts homologated to stay ahead of circuit rivals Mercedes-Benz and Audi, with each increment resulting in a further run of road cars—the familiar Evolution, Evolution II, and Sport Evolution beloved of enthusiasts and collectors.

By the time the factory bowed out at the end of the 1991 season, the M3 engine had grown to 2.5 liters and, in race trim, was giving close on 320 horsepower at upward of 9,000 rpm. Even without factory backing, M3s continued their winning streak for many more years in the hands of private teams and individual entrants. This original M3 was one of those rare cars to become a legend in its own lifetime, both on the racetrack and on the road—and even now, as each new production M car comes out, there are calls from the enthusiast community for BMW to rekindle the spirit of that explosive, exhilarating first one.

Giving it the works: The M3 engine used the M10 block paired with a sixteen-valve DOHC head derived from that of the original M5's six. Early versions gave a punchy 200 hp, but race cars were good for over 320 hp at 9,000 rpm.

The innocuously labeled 320iS was an M3 in all but name. M engineers developed it with a 2.0-liter engine for the Italian market tax break—and it was rated at 190 hp, scarcely down on the original.

The M3 was offered in Cabriolet form two years after the original launched. It proved to be a true sports car to drive, with little sacrifice to the model's legendary tautness and lightning responses. Even so, it accounted for fewer than one in twenty M3 sales.

BMW was slow to market with a Touring wagon version of the 3-Series, but it quickly caught on with the fashionable style-conscious country set, despite the narrow tailgate and limited capacity of its cargo bay.

Familiar face: The four-lamp frontal signature of the second 3-Series was universally recognized and became the calling card for almost every BMW since. Echoes of this classic look are still referenced in the latest models, five decades later.

The M3's sixteen-valve, dual overhead-cam engine was systematically developed from its original 200 horsepower to more than 320 in racing form.

(CONTINUED FROM PAGE 67)

raised the 3's profile to a global audience, but even before the end of the mainstream model's tenure, some cracks were beginning to show. The sedan was beginning to look upright and old-fashioned, the tight rear-seat accommodation was an increasing source of complaint, and rivals were already circling like vultures—Mercedes with its "small" 190, Audi with its svelte new third-generation 80 offering Quattro four-wheel drive, and even the occasional flourish from Alfa Romeo.

In short, the E30 generation—especially the boxy-looking sedan—had stayed too long, and it was time to bow out gracefully. The more fashionable Cabrio and Touring enjoyed a long and successful dotage before finally handing over the reins to their E36 successors. That they were able to last that long and still seem at the top of their game is a tribute to the clear quality of the original design.

COLLECTABILITY

Collectable? The one-word answer is "totally." The M3 has always been in demand, and the more that twenty-first-century M-cars turn to turbos, automatic transmissions, all-wheel drive, and computer-controlled chassis dynamics, the greater the clamor for the purist, totally unfiltered raw thrills of the original race-derived M3 becomes. But more recently, the focus has expanded to the entire E30 lineup, with more and more people recognizing its deep-down quality, the integrity of its engineering, and the fun factor that every model is capable of delivering. The classy Cabriolets are particularly sought after.

Standout model: M3 Sport Evolution. This is the crème de la crème of M3s and probably the only modern BMW to sell well into six-digit-dollar-values. But as M3 demand ratchets up alarmingly, with collectors rushing to grab their own examples, there are signs of a price bubble in danger of bursting. Worth the risk? For the M-car that started it all—you can't get any more original than that—it's a steep entry ticket but quite possibly a risk worth taking.

Hidden gem: 318iS. There are plenty of gems within the E30 ranks, but this is one of the best-kept secrets. Outwardly simple and standard-looking, it is armed with a deliciously revvy sixteen-valve version of the M10 engine, and with 136 horsepower, it makes for an enticing drive.

Wild card: 325iX. Not many people noticed—let alone bought—this all-wheel-drive version of the silken 170-horsepower, six-cylinder 3 when it came out in '85, and it wasn't as engaging as its rear-wheel parent to drive. But maybe one day that rarity and uniqueness could make it a curiosity to covet.

Standout model: M3 Sport Evolution
What other standout model could there possibly be than the fastest and most powerful—at 238 hp—of all? It's sensational, it's the ultimate M icon—and only six hundred were built. That means it's one of the most sought-after BMWs by the wealthiest collectors, and six-figure prices prove the point.

Wild card: 325iX
Wild card or simply an intriguing choice? The four-wheel drive 325iX, also known as the 325i Allrad, had an intriguingly complex driveline that was great for snow but didn't quite translate into Delta Integrale–type thrills the rest of the year.

BMW 3-SERIES E30: POWER TRAINS AND PERFORMANCE

Model	316i	318i	318iS	320i	320iS	M3	323i	324d	324td	325i
Engine Code	M10	M10	M40	M20	S14	S14	M20	M21D24	M21D24	M20
Configuration	Inline 4, SOHC 8V	Inline 4, SOHC 8V	Inline 4, DOHC, 16V	Inline 6, SOHC 12V	Inline 4, DOHC 16V	Inline 4, DOHC 16V	Inline 6, SOHC 12V	Inline 6, SOHC 12V	Inline 6, SOHC 12V	Inline 6, SOHC 12V
Capacity (cc)	1,766	1,754	1,796	1,990	1,990	2,302	2,315	2,443	2,443	2,495
Fueling	Fuel injection	Fuel injection	Fuel injection	Fuel injection	Fuel injection	Fuel injection	Fuel injection	Diesel	Turbo diesel	Fuel injection
Power @ rpm	102 @ 6,000	105 @ 5,800	136 @ 6,000	125–129 @ 5,800	190 @ 6,900	195 @ 6,750	139–150 @ 6,000	86 @ 4,600	115 @ 4,800	171 @ 5,800
Torque @ rpm (N m)	137 @ 4,000	145 @ 4,500	169 @ 4,600	171 @ 4,000	2110 @ 4,900	226 @ 4,600	205 @ 4,000	149 @ 2,500	210 @ 2,400	225 @ 4,000
Transmissions	M4, later M5	M5	M5	M5, AT4	M5	M5	M5	M5	M5, A4	M5, A4
Max Speed (km/h)	175	185	205	200	227	235	205	170	185	220
0–100 km/h	12.0	11.5	10.0	10.0	7.9	7.0	9.0	16.0	12.5	8.0
Comments	Early 316 had 90 hp		Late addition in 1990		Limited-edition M3 for Italy, Portugal	Later-evolution models take power to 238 hp				

Of the twenty BMW Art Cars commissioned to date, five have been painted on the canvas of a 3-Series. Ken Done's Group A racing E30 M3 from 1989 was the eighth in the Art Car sequence and was unveiled at the same time as fellow Australian Michael Jagamara's similar model. Done's vivid color scheme depicts the vibrancy of Australia's nature and climate, and the car enjoyed a high-profile racing career before becoming a much-admired museum piece.

THE BIG LEAGUE BECKONS

Third time lucky: The E36 generation took over from a very successful predecessor and marked the first big change in direction—notably the quick introduction of the coupe derivative, here in M3 form.

Toward the end of the 1980s, BMW was feeling pretty pleased with itself—and with good reason. It had won a coveted Formula One World Championship with Nelson Piquet in a BMW-powered Brabham, it had comprehensively out-engineered archrival Mercedes thanks to the spectacular V-12 engine in its elegant new 7-Series, and the smaller—and equally svelte—third-generation 5-Series was hitting Mercedes where it hurt most, in the profitable luxury E-segment. BMW knew, too, that a smart new 3-Series was on the starting blocks and ready to launch into battle in the big-numbers medium sector.

The company was well accustomed to fulsome praise in both the financial press and the consumer magazines. It wasn't so familiar with criticism—especially the often strident criticism that greeted the unveiling of the third-generation 3-Series, code-named E36.

The issue wasn't the engineering, or the specifications, or even the price. It was much starker than that: It was simply "that front" that was stirring up the emotion among BMW loyalists.

Today, with several decades' hindsight, it is hard to see what all the fuss was about. But at the time it was strongly felt and equally strongly expressed—and if the furor did show something useful, it was that some customers and commentators were finding it hard to accept change. It revealed how strongly invested people can be in their own perception of a brand, and how they resist any changes to that familiar identity. For BMW, it must have been both galling and reassuring at the same time to see that its customers had such a powerful image of the brand, something that would make it harder for them to move on.

Many automakers, especially those with strong brand identities, come up against this precise problem. BMW would hit it again big-time a decade later when the E65 7-Series, first of the models designed under Chris Bangle's watch, launched to massive media-wide opprobrium and even mockery in 2001.

But back to the genesis of the E36 3-Series. Engineering priorities had forced the outgoing E30 generation to be more cautious than its designers originally wanted; its cues were strongly linked to those of the initial E21, so in effect, the outgoing car actually harked back to 1975, when the E21 was launched. And this was precisely the image of BMW that those critics were trying to cling on to. The designers knew this, yet they reasoned that customers had had plenty of time to get used to the new-look 7- and 5-Series, with their smooth, flowing lines and gentle, rounded fronts having banished the outdated boxy, slab-sided shapes.

No such luck—the panic was intense. Thankfully, however, it was short-lived. The smooth new shape came across as the very opposite of boxy. The E36 showed a strong wedge profile, with the fast backlight and high deck line diluting the familiar three-box profile. The sides were gently rounded in profile, the side windows were much less upright, and the nose, instead of standing upright or being angled forward, was canted rearward—including the grille and headlights, which now swept smoothly backward.

Evolution in action: Clearly visible in side view is how the third 3-Series brought in a smoother, more flowing wedge profile, along with a more hatch-like fastback tail, even though both trunk volume and passenger space were increased.

For the first time, BMW produced a bespoke coupe version, which the firm boasted shared no panels at all with the sedan. Noticeable is how the distinct coupe style retains more of a notchback feel than the sedan.

BMW 3-SERIES E36: TIMELINE

1990	Four-door sedan range launched: 316i, 318i, 320i, and 325i; latter are both 6-cylinder
1991	Two-door Coupe models added: 325i, later 320i, 318iS
1991	6-cylinder diesel 325td and 325tds added
1992	VANOS variable valve timing added to 6-cylinder gasoline engines
1993	Cabriolet versions added, initially 318iS and 325i
1993	M3 Coupe launched, with Motorsport-developed 24-valve 6 giving 286 hp
1994	Two-door Compact hatch launched, initially as 316i; 4-cylinder; 318tds diesel added
1994	M3 sedan and Cabrio versions added
1995	Touring wagons added, with similar engine options to sedan; 328i replaces 325i
1996	M3 revised with enlarged 320-hp engine (Coupe and Cabriolet only)
1997	Compact gets sporting 323i version with 2.5-liter 6; 318ti Compact added with 16-valve, 1.8-liter engine
1998	Sedan models replaced by new E46 generation
1999	Coupe, Cabriolet, and Touring models phased out
2000	Compact phased out

BMW 3-SERIES E36: STYLES AND VERSIONS

Body Style	4-Door Sedan	2-Door Coupe	2-Door Cabriolet	4-Door Touring	3-Door Compact	Grand Total, All Models
Length (mm)	4,435	4,435	4,435	4,435	4,210	
Wheel-base (mm)	2,700	2,700	2,700	2,700	2,700	
Curb Weight	1,185–1,350	1,160–1,300	1,325–1,400	1,285–1,400	1,175–1,255	
Number Built	1,552,567	468,492	184,430	130,611	396,162	2,732,262

BMW 3-SERIES E36: BASICS AND ENGINEERING

Layout	Front engine, rear-wheel drive
Suspension F/R	MacPherson struts/Z-axle (except Compact)
Steering	Rack and pinion, power
Braking	Disc
Comments	Compact used carryover rear suspension from E30 model

DESIGN UP CLOSE E36

The E36 represents an important step change for BMW in the medium segment: the marque's transition from the straight-line orthodoxy of the 1960s and 1970s to a more flowing design language as the turn of the millennium approached. Design chief Claus Luthe skillfully mentored the evolution of the corporate identity, sweeping the lights backward into the hood, taking with them the widened BMW kidney grilles (now framed by body-color moldings) to form a smooth ensemble. This was aided by the glass coverings for the quad headlights, but it proved to be something of a flashpoint among BMW purists at the time.

With raised hood and rear deck lines and faster front and rear screens, the sedan's side profile became more hatchback-like, distancing this generation from its three-box predecessors. However, that older typology was still referenced by the new Coupe version, which had a visually separated trunk area, while the Touring and Cabrio derivatives continued with the successful proportions and stance of their E30 forebears, giving further continuity to an already developing premium identity for the brand.

As is so often the case with BMW, the only facelift actions during the life of these models were so subtle as to be virtually undetectable to non owners. This served yet again to boost brand value and help prevent existing models appearing outdated.

What was all the fuss about? Back in 1990 purists fretted about the new 3-Series' transparent headlamp covers. Today, it can be seen as aerodynamic, skillfully ahead of its time—and Claus Luthe's work is held in high regard.

Any designer who tampered with the traditional BMW grille risked vilification in the press, and the E36's rear-slanted treatment, with its subtly projecting body-color surround, wasn't welcomed at the time—but the commotion soon died down.

Three of a kind—at least mechanically. For the first time, 3-Series versions were structurally differentiated, with the Cabriolet and Coupe lower and wider-tracked than the sedan (rear). Very few body components were shared.

The main point of contention: The quadruple headlights were now faired in behind flush transparent covers. The openly expressed fear was that by hiding the lights behind covers, a key BMW brand hallmark would be lost and the model would look like any other midsize car on the market.

THE STORM PASSES

BMW managers—and, it has to be said, the whole design community—stood their ground, arguing that the 3-Series needed to move on, to transition into the new era of streamlining and aerodynamic efficiency.

Before long, the storm had passed, and it was time to consider the E36 more objectively. In addition to the new shape, there was much to be excited about: The wheelbase was stretched, promising much-needed extra legroom in the back seats. The same range of highly praised engines was offered, now all with fuel injection and five-speed transmissions. And a sophisticated new rear suspension layout—the so-called Z-axle from the advanced Z1 sports car—hinted at even better handling and ride.

Inside, the ambience was much friendlier than the stiff, sit-up-and-beg formality of the E30. The colors were paler, the steering wheel and instrument housing were smaller and neater, and the lower seating position and higher beltline made it feel cozier and more coupe-like. But sitting in the rear seat brought some less welcome revelations. Not much of the wheelbase stretch had actually been allocated to improving legroom, which still felt tight, and the effect was amplified by the hard plastic backs to the front seats. Indeed, compared with the previous car, there was much more plastic on view, and the perceived quality seemed to have suffered. Up front for the driver, though, things were notably better. The engines, as might be expected, were every bit as eager as before, the steering was quicker and more communicative, and the new rear suspension made for a smoother ride and better stability. BMW was quick to address the interior quality niggles, too.

For the first year, it was just the four-door sedan, with a choice of engines ranging from the 316i, with 100 horsepower, to the six-cylinder 325i, with almost double. It sold faster than any previous BMW, shifting more than a quarter million units in its first year. Second-cheapest, the 318i was easily the favorite. The omens were looking good.

What was never in doubt with the third-generation: Three was the excellence of the driving experience. Thanks to the elaborate Z-axle rear suspension, the models' ride comfort, steering, and handling always scored highly in reviews.

THE COUPE OFFERS A TWO-DOOR OPTION

It had been done before (remember the Morris Marina?), but BMW did it with particular aplomb. Summer 1991 saw the launch of the first 3-Series Coupe, which served as the two-door alternative—not the cheaper option, but more stylish, more intimate, and more individual. Design director Claus Luthe took great pride in explaining that every single body panel was unique to the Coupe; nothing was shared with the sedan, though of course the underpinnings and the overall engineering approach were the same.

This freedom allowed Luthe to make the Coupe version lower than the sedan, with more steeply raked screens and a more notchback rear to echo the still-revered three-box 2002. The doors had frameless windows to allow a sleeker roofline, and an innovation in the segment was the use of the big 850i's mechanism, which raised the side glass a few millimeters after the door was shut, ensuring a good seal once closed.

The Coupe had firmer suspension settings than the sedan, and its initial six-cylinder 320i and 325i engines were later joined by the 1.8-liter, sixteen-cylinder 318iS four, which soon took over the bestseller slot. By now, most of the quality quibbles had been ironed out, the cars were good to drive on all types of roads, and there was a general consensus that these Coupes represented BMW at its very best.

EXPANDING THE OFFER

With the Coupe range selling well, BMW seemed in no hurry to introduce further body styles, and it was a full two years into the E36's production run before the Cabriolet version appeared. Initially led by the six-cylinder 320i and 325i, the Cabrio was a typically thorough and stylish design. Though the structure was based on the Coupe, the engine hood was different, and windshield and A-pillars were moved back slightly, reportedly as part of the quest for a more flattering side profile. And to avoid the need for an unsightly fixed roll bar, BMW introduced pyrotechnic pop-up bars behind the rear seats that deployed if an accident looked unavoidable.

The impeccably finished Cabrio was always the priciest of the 3-Series models—apart from the M3, of course—but that did not hamper sales. It notched up consistent five-figure annual totals right through until production run-out at the end of the decade.

Despite its fastback allure, the core 3-Series was always a sedan with a separate trunk. For hatchback versality buyers had to either trade up to a Touring wagon or save cash on a three-door Compact. This is the entry-level 316i: later, a hot six-cylinder 323i and even diesels were added.

The Touring station wagon was even slower to arrive, first appearing in 1995 with the full suite of engines apart from the lowly 1.6. Though the Touring body was no greater in length than the other models, it made the 3-Series look bigger and more substantial. It now had a much roomier cargo area, with minimal suspension intrusion thanks to the clever and compact Z-axle, and the taillight clusters no longer restricted the loading width. With this generation change, the Touring had become a proper wagon for family holiday travel rather than a stylish runabout best suited to a short weekend leisure break.

A COMPACT SURPRISE— AND A VERY BIG ONE

BMW top brass had for some time been keeping their eyes open for possible threats to the company's position, not just from familiar premium players such as Mercedes and Audi, but elsewhere, too. What had caught the directors' eyes was the Mark 3 Volkswagen Golf—not the car itself, which had been given a massive roasting in the press, but the fact that VW proposed to offer it with a clever 2.8-liter, six-cylinder engine as a luxury option. This, to BMW, signaled an aspiration to climb upmarket and possibly threaten the lower reaches of the 3-Series range. There was a parallel concern that top hot hatchbacks such as the Golf GTI could undercut BMW prices to entice its sports-minded customers away.

The surprise that BMW sprang on the industry in September 1993 was the Compact, a shortened and lightened version of the 3-Series, but with 22 centimeters lopped off the trunk to create a chubby three-door hatchback. Cheaper and simpler than the donor sedan, it promised to be lighter, more agile, and more fun—handy qualities in the fight against the sporting Golfs. Yet though it bore the BMW badge, closer examination suggested that it had been built down to a price, very much not usual BMW practice. The dashboard, for instance, was a straight graft from the previous E30 generation, and even some of the switches were ancient hand-me-downs of the push-pull type last seen on VW Beetles. But BMW's biggest faux pas was to save cash by borrowing the outdated previous-generation rear suspension from the E30, complete with its propensity to slide outward on curves taken too enthusiastically.

Still, all this made the Compact quite fun to drive on twisty roads in good conditions, and later versions with

(CONTINUED ON PAGE 84)

MOTORSPORT AND THE M3

Just as the regular E36 models marked a big shift away from the flavor of their predecessors, so, too, did the new M3—but in a very different way. The original M3 had been unashamedly vibrant and energetic, a barely disguised racing car with few concessions to comfort or refinement. Its replacement was just the opposite: a smooth coupe with a sophisticated 3.0-liter six-cylinder engine instead of a highly strung four. At first the purists, who'd been hoping for another raw and ready racer, were horrified. But that was before they tried the new car.

The new S50B0 engine, specially developed for the car from the small-block twenty-four-valve six used in the standard 3- and 5-Series, was something astonishing and truly remarkable. It seemed to have everything: great flexibility, a punchy midrange, and searing power right through to its strident 7,250 rpm redline. It was fabulously smooth and refined, too, but still truly came on song when pushed harder.

One of the secrets of the S50 was its introduction of a system developed by BMW Motorsport: VANOS variable valve timing, initially only for the inlet side of the engine. This allowed the breathing to be optimized for efficient low-down torque but then to adjust itself to the very different timings needed for high rpm and peak power. Allied to the 286-horsepower powerhouse was a new five-speed transmission (without the dogleg pattern, to many people's relief) and a limited-slip differential on the rear axle. Steering, suspension, dampers, and brakes were all upgraded by Motorsport engineers to near-race specifications.

Bodywork changes—aside from the distinctive wider alloy wheels—were subtle and not immediately obvious. There wasn't any rear wing, but there was a bigger frontal air dam, different sill moldings, and a rear apron incorporating a diffuser slot. The ride height was lower, too. Inside, the M3 was plushly kitted out, as befitted its DM 80,000 launch price—almost double the cost of a regular 318iS at the time.

No two ways about it, this M3 was a sensation. Though some

M3 take two: The second-generation M3 raised its game with a fabulous twenty-four-valve six of prodigious power and refinement. But M-hardliners lamented the loss of the raw, racerish edginess of the original, fearing BMW had gone soft.

commentators still clung on to their memories of the hyperactive, race-focused original, bemoaning the new M3's extra weight and less sensitive steering, the overwhelming consensus was that the new car was something genuinely special—a sophisticated grand tourer that was easy to drive but packed a prodigious punch when needed. Before long, a Cabrio version was added, with power top standard; later, as something of a surprise, came a four-door sedan.

As had happened too many times before, North American customers were short-changed. When the M3 belatedly arrived stateside, its engine was cruelly throttled back to just 240 horsepower and 6,000 rpm because that was all that the optional automatic transmission could handle. And when the so-called Evolution upgrade came in 1995, it brought a larger 3.2-liter engine and 321 horsepower for Europe, but just 243 for the US. That European engine, now with VANOS on the exhaust side, too, marked yet another high in BMW's engineering, and it was paired with a six-speed gearbox—though, again, never for the US. This model also marked the debut of a BMW feature that few people took to: the sequential manual gearbox (SMG) automated manual shift. Thankfully only an option, it came across as clunky and obstructive, and it went through several more iterations and software fixes before it became acceptable.

Several limited editions distinguished different versions of the M3, but in contrast to the original M3, these were in general not aimed at homologating innovations for the racetrack—with the exception of the final GT, a 400-run lightened special edition aimed at GT Class II and IMSA GT. In Europe, BMW's motorsport plans for the M3 had been scuppered when new rules were introduced in 1993 for the principal Super Touring Car series that limited engine capacity to 2.0 liters, which excluded the M3. The car was, however, dominant in Group N races for near-standard cars and a regular winner of 24-hour events at legendary tracks such as Spa and the Nürburgring. BMW was still fielding the 318iS in Super Touring events and continued selling M3 race kits; these helped the new M3s take many national championship titles, even in hillclimb series.

Even the diesel version of the 3-Series shone in some events. Remarkably, a 320d turbodiesel, with 200 horsepower on tap, took outright victory in the 1988 Nürburgring 24 Hours, the first time a diesel car had ever won the event.

In North America, the picture was very different. BMW encouraged competition by offering stripped-out Lightweight M3s aimed at teams and private entrants. The cars came with a kit of race parts in the trunk, ready to fit—and they quickly became successful in the GT classes in a variety of sports car races and championships, including several class and overall manufacturers' titles. The M3's winning streak continued unabated until the end of the decade and the arrival of the next-generation E46 M3—and that would have its own dramas as BMW found clever ways to beat the Porsche 911s.

An E36 M3 stops for tires and fuel in a 24-hour race: With its larger and more potent engine, the six-cylinder M3 was no longer eligible for the European Super Touring series but became a dominant force in near-standard Group-N long-distance events.

Born in the USA: The 318i has the distinction of being the first BMW to be built in North America. This white example was signed by every employee at the South Carolina manufacturing complex, which has since become BMW's biggest plant worldwide.

(CONTINUED FROM PAGE 81)

the more potent 1.8-liter, sixteen-valve (318iS) and 2.5-liter, twenty-four-valve six (323i) were quicker and did provide a distant echo of the 2002's fun factor. When the numbers were totted up at the end of the Compact's seven-year production run, almost four hundred thousand had found buyers. Not a shabby performance, for sure, and in all probability a decent return on the small investment it had required. But had it been worth the early-days damage to BMW's precious reputation? Because overall, in stark contrast to its svelte Coupe and Cabrio peers, the Compact had failed to represent BMW at its best.

SHOCK TAKEOVER

As for the bigger surprise, it occurred on the last day of January 1994, just four months after the Compact's unveiling and mere weeks ahead of its international driving launch. In a sudden announcement that stunned the automotive and business worlds, BMW stated that it had reached an agreement to buy the British Rover Group lock, stock, and barrel. As part of the £800 million deal, BMW would take ownership of the Rover, MG, and Mini brands, countless factories scattered across the UK, a clutch of legacy brand names such as Triumph, plus the most coveted prize of all: the Land Rover brand and the rights to its exclusive Range Rover nameplate.

What does all this have to do with the Compact? Nothing technically, but everything commercially. The Compact had been BMW's line of defense against incursions into its territory by upwardly mobile volume players like VW. But with the acquisition of Rover, BMW could now jump straight into its long-term strategy of fostering its in-house mass-market brand to give the combined group what it needed most—much higher volumes to provide the economies of scale and the lower industrial costs necessary to compete against global players. If BMW were simply to pump those massive extra volumes through its own brand, reasoned the directors, that would risk diluting the exclusivity of the BMW nameplate. Channeling those numbers instead through the Mini and Rover badges would keep the parent company protected while generating the big volumes that the cost accountants prescribed.

The lower-cost Compact hatchback was announced just months before the Rover takeover. It proved a useful asset during the long wait for fresh midsized Rover models.

BMW 316i COMPACT.
JE GRÖSSER DIE KAUFLUST,
DESTO BESSER IST ER.

Die Situation: Kurzer Samstag in der Stadt. Viel zu erledigen, aber viel zu wenig Zeit. Trotzdem schnell noch mal in das neue Modegeschäft. Zwei Outfits, drei Paar Schuhe, Tüten so groß wie Reisekoffer.

Der Grundsatz: Ein Automobil für die Stadt muß nicht nur kompakt und wendig sein – es soll darüber hinaus noch Stauraum bieten wie ein Großer und das Be- und Entladen so einfach wie irgend möglich machen.

Die technische Lösung: Eine niedrige Ladekante mit weit zu öffnender Heckklappe. Beide Rücksitze sind getrennt umklappbar und schaffen im Handumdrehen erstaunliche eintausend Liter Laderaum.

Weitere Informationen unter Tel. 01 30/33 35, Fax 0 89/4 70 81 62, BTX ∗ BMW #

BMW
FREUDE AM FAHREN

The tagline reads: "The more you love shopping, the better this car gets." With the launch of the shorter, handier, and cheaper Compact hatchback, BMW began targeting a fashionable urban audience.

With Rover now on board, the Compact appeared to have been made redundant at the very moment of its launch. Top BMW officials at the Compact's presentation openly admitted as much, with some even saying that the model might not have been given the green light had the company known that it would scoop up the ailing British group. Of course, with just weeks between the takeover announcement and the Compact's market release, it was far too late to pull the plug.

Yet as the events that unfolded over the next months and years would so starkly highlight, Rover proved such a problem child that it became an all-enveloping disaster rather than a strategic asset—and BMW had cause to celebrate its low-stakes, entry-level Compact after all.

The E36 offered a particularly neat, low-set driving position, which contrasted with the more upright feel of its forebear. The steering wheel felt surprisingly large, though, and early cars had some trim quality issues, soon resolved.

COLLECTABILITY

The E36 is one of those neglected interim 3-Series generations sandwiched between more glamorous earlier and later designs, but interest is picking up as it enters classic territory. All versions apart from the lowliest 1.8-liter diesel are light, agile, and fun to drive thanks to the clever Z-axle rear suspension and eager engines. Once again, the Coupe and Cabriolet provide collectible highlights, but choose carefully—many may have been run by multiple owners whose aspirations stretched further than their maintenance budgets.

Standout model: M3 Evolution. Yet again the M3 swings it as the headline choice. Many astute collectors are now looking to this generation for better value than the inflated E30 and the later E46, whose M3 prices are being pushed up by the frenzy surrounding the exclusive CSL. And besides, this E36 M3 offers a breathtaking drive, with more raw energy than its successor.

Hidden gem: 318iS. Yes, the same package as with the E30. The six-cylinder models may offer better power and speed, but the lighter, simpler, and more agile sixteen-valve 4 makes for easier fun—and lower bills.

Wild card: 323i Compact. This chopped-down, low-cost 3 was hammered by the press on its launch in 1993, and most survivors are likely neglected clunkers on their last legs. But the rare six-cylinder 323i could be a great find if not too molested, and the fuel-injection 318ti could be fun, too.

Standout model: M3 GT
Early concerns about the M3 having gone soft in six-cylinder form soon dissipated, and the 1996-on evolution models, with their cracking 3.2-liter engine and six-speed box, are a top choice. Especially this four-hundred-run GT limited edition, all finished in British racing green.

Hidden gem: 318iS
No apologies for including the 318iS as a hidden gem. The sixteen-valve engine, with an eager 140 hp, is sweetness itself and blends beautifully with a chassis that is eager to please. And the bills will be lower, too.

TOP AND LEFT: Not so much a car, but more a research project that BMW was pressured into putting into production: The Z1 sports car's mission was to trial a new form of construction, with a galvanized steel chassis topped by bolt-on composite panels. Such was the clamor on its unveiling at the 1987 Frankfurt show that orders flooded in and BMW felt obliged to build it. Yet when those cars arrived they looked cute but proved disappointing: They were expensive, heavy, and underpowered, despite the 325i engine and clever suspension. Just eight thousand were made—but now they are regarded as collectors' classics.

MORE PERMUTATIONS

With a full set of engines and a suite of five body styles at its disposal, BMW soon began playing with different permutations to help maximize the 3-Series' market coverage. But company planners knew the limits of acceptability and when to stop playing mix-and-match—while the Touring wagon was available with almost every engine, the Compact never got the top six-cylinder engines, and there were never any diesel Coupes or Cabrios. Even the exclusive, super-powerful M3 engine spread from the Coupe to the Cabrio and sedan, but never to the Compact—except as a one-off special.

Most of these combinations worked well, and even the sweet 1.6-liter motor in the Coupe and heavier Touring gave reasonable performance. But while the six-cylinder diesels were acceptably lively and quite refined for their time, the same could not be said for their 1.8-liter, four-cylinder counterpart. This feeble unit never felt like its claimed 90 horsepower and struggled to impart momentum into any of its host models, even the lighter Compact—a rare case of BMW failing to deliver on its fun-to-drive promise.

The Touring station wagon version of the E36 3-Series offered a useful step-up in roominess and versatility over its predecessor, with the broader tailgate especially appreciated.

CLOSING BALANCE: E36 3-SERIES

While "soldiered on" is the phrase often employed to describe the run-out phase of volume-market models, it hardly seems appropriate for the E36 3-Series. Sedan versions stayed in full production until early 1999, when the replacement E46 first drew breath, and more than 1.5 million found buyers worldwide in its eight-year reign. The Compact, much criticized in its early months, stayed into the new millennium and had a positive balance, too, finding almost four hundred thousand fans in its shorter lifetime. It demonstrated a healthy consumer demand for a more accessible BMW that, while not necessarily delivering on the brand promise of sparkling performance, certainly ticked enough quality and premium-appeal boxes to make it a commercial success. And, less charitably, its success could also have reflected BMW management's failure to get to grips with its Rover acquisition and produce a British-branded model capable of protecting the lower reaches of the German range.

As for the Coupe, Cabrio, and Touring variants, they stayed in full swing for a good year more, with their respective totals of 468,000, 184,000, and 130,000 bringing the overall tally for this generation to more than 2.7 million—BMW's biggest hit to date, and a ringing endorsement of its strategy to extend its premium offer laterally into additional sectors. And on a broader level, the smooth handover from one generation to the next, with all models remaining desirable (and not subjected to damaging discounts) until their run-out dates, showed the shrewdness of BMW management's product planning. For competing manufacturers, this provided a handy lesson in reinforcing brand equity, supporting the values of owners' own vehicles, and feeding the virtuous circle of promoting consumer demand for the replacement models.

The nine-year lifespan of this third-generation 3-Series neatly brackets a period of great change for BMW, which saw spectacular expansion and growing influence but also the painful internal upheavals following the strategic realignments that resulted from the Rover acquisition. For the broader auto industry, the success of this generation showed, for perhaps the very first time, that with skillful management it was indeed possible to produce a premium car in mass-market numbers—without damaging the value of the brand or its perceived exclusivity. This lesson, demonstrated so eloquently with the E36 generation, would be the key to BMW's climb to the very top of the premium tree little more than a decade later.

BMW 3-SERIES E36: POWER TRAINS AND PERFORMANCE

Model	316i	318i	318iS	320i	323ti Compact	325i	328i	318tds	325td	325tds	M3
Engine Code	M40	M40	M42/ M44	M50/M52	M50/M52	M52	M52	M41	M51	M51	S50B30/ B32
Configuration	Inline 4, SOHC 8V	Inline 4, SOHC 8V	Inline 4, DOHC 16V	Inline 6, DOHC 24V	Inline 6, DOHC 24V	Inline 6, DOHC 24V	Inline 6, DOHC 24V	Inline 4, SOHC 8V turbo intercooled diesel	Inline 6, SOHC 12V turbo diesel	Inline 6, SOHC 12V turbo inter-cooled diesel	Inline 6, DOHC 24V, single VANOS VVT; double on B32
Capacity (cc)	1,596	1,796	1,796	1,991	2,494	2,494	2,793	1,665	2,497	2,497	2,990/ 3,201
Fueling	Fuel injection	Fuel injection	Fuel injection	Fuel injection	Fuel injection	Fuel injection	Fuel injection	Diesel injection	Diesel injection	Diesel injection	Fuel injection
Power @ rpm	100/102 @ 5,500	113 @ 5,500	140 @ 6,000	150 @ 5,900	170 @ 5,500	192 @ 5,900	193 @ 5,300	90 @ 4,400	115 @ 4,800	143 @ 4,800	286 @ 7,000/ 321 @ 7,400
Torque @ rpm (N m)	141 @ 4,250	162 @ 4,250	175 @ 4,500	190 @ 4,200	245 @ 3,950	245 @ 4,200	280 @ 3,950	190 @ 2,200	222 @ 1,900	260 @ 1,900	320 @ 3,600/ 350 @ 3,250
Transmissions	M5/4AT	M5/4AT	M5/4AT	M5/5AT	M5/4AT	M5/5AT	M5/5AT	M5	M5/4AT	M5/5AT	M5/M6
Max Speed (km/h)	191	198	213	214	230	233	236	182	198	210	250 (limited)
0–100 km/h	13.1	11.5	10.2	9.8	7.8	8.0	7.3	14.4	12.0	10.0	6.0/5.5
Comments			1,895cc M44 from 1995	M52 VANOS VVT on 6-cylinder engines from 1992							US versions: 240 hp @ 6,000 rpm and 243 on B32 from 1995

The six-cylinder M3 was a big success in all markets, with the evolution version an even more engaging drive. But pity the poor North American customers: They only got 240 or 243 hp, compared with 286 and a searing 321 for European buyers.

Playing games with names: The badge reads 323i, but the engine is actually the creamy 2.5-liter six. BMW planners made the move to distance this version from the newly introduced 328i flagship model.

M LB 300

PEAK PERFORMANCE: 3-SERIES HITS NEW HEIGHTS

Familiar but enhanced in every detail: In a classic BMW tactic, the new E46 shape earned instant recognition even though it was actually a complete restyle and both longer and wider than before.

Top of the E46's engineering brief were a big step-up in safety and more passenger space. BMW did well to keep the weight increase to under 100 kg, but suspension refinements gave a much smoother ride with no sacrifice in agility.

From the initial euphoria of the Rover acquisition in 1994 to the humiliating denouement of its disposal six exhausting years later, the nightmare of the so-called English Patient took a huge toll on BMW—not on its cars, but on corporate morale, on group finances, and on the executives who burned themselves out or parted company with the group trying to cope. Yet to their great credit, those doing the actual, practical work did not flinch. In spite of the turmoil going on around them, the planners, the engineers, the designers, and the test teams continued their work unruffled. In fact, some of the most admired models in BMW's entire history were developed during this period.

The distraction of Rover aside, the core BMW business in the mid-'90s was in excellent shape. Sales were booming on all key model lines, the US plant in Spartanburg, North Carolina, was churning out the Z3 sports car, and the shiny new FIZ research center back home in Munich had been expanded to more closely integrate all the functions of vehicle creation—research, design, engineering, development, and product planning. There was plenty of cash in the bank, and the omens were fair.

The Geneva show of March 1998 was positively triumphal for the core BMW marque, with the launch of the tremendous new M5—the first sports sedan with more than 400 horsepower—and the unveiling of the next generation of its bestseller, the 3-Series.

Superficially, the E46 generation was everything that people had come to expect from a new BMW: an incremental fine-tuning of a recent recipe that had proven spectacularly successful over eight years, a suite of engines already acknowledged to be the best in the business, and a major step-up in safety provision, both structurally and in terms of electronic equipment. It didn't matter that the new car, though even neater and more harmonious than the one it replaced, carried exactly the same aura as before. The differences were sufficient for those in the know to spot them—and that's what seemed to count.

BMW 3-SERIES E46: TIMELINE

1998	E46 generation launched as 316i (105 hp) and 318i (118 hp) four-door sedans; 320d diesel (initially 136 hp) and 6-cylinder gasoline 320i (150 hp), 323i (2.5 liters, 170 hp), and 328i (193 hp) follow.
1999	Ci Coupe version revealed at Geneva show and goes on sale late 1999; Touring wagon versions introduced
2000	320i models have enlarged 2.2-liter engine giving 170 hp; 6-cylinder 330d with 184 hp added; Cabriolet and four-wheel drive xDrive models introduced
2000	New M3 Coupe has 3.2-liter double VANOS 6-cylinder engine (343 hp) and 6-speed Getrag gearbox or optional SMG sequential shift
2001	World-first variable valve timing and lift (Valvetronic) on 4-cylinder gasoline models: 316i (1.8 liters, 115 hp) and 318i (1.9 liters, 140 hp)
2001	ti Compact models, with shortened hatchback bodywork, added
2002	Diesel engines and 6-speed transmission made available across all body styles
2003	Race-inspired lightweight M3 CSL launched, with 360 hp, aluminum front panels, and carbon-fiber roof panel
2004	Sedan versions replaced by new E90 series
2006	Other derivatives phased out

NEW DESIGN PHILOSOPHY? NOT QUITE YET

The E46 had a smoothly rounded front and a raised center section on the engine hood that wrapped harmoniously around the front edge to sweep down into the grille surround; the body sides were perfectly profiled, with unbroken shoulder, waist, and feature lines. This was hardly the stuff of revolution. Yet just as the E46 program was about to kick off, a new face had joined the design team, to end the interregnum following Claus Luthe's departure in 1990. Chris Bangle, an American, was the first non-European to lead BMW design, and among his early comments on arrival

BMW 3-SERIES E46: BASICS AND ENGINEERING

Layout	Front engine, rear-wheel drive/all-wheel drive
Suspension F/R	MacPherson struts, lower wishbones/ multilink Z-axle
Steering	Rack and pinion, power assisted
Braking	Discs, ABS

BMW 3-SERIES E46: STYLES AND VERSIONS

Body Style	4-Door Sedan	2-Door Coupe	2-Door Cabriolet	4-Door Touring	3-Door Compact	M3	Grand Total, All Models
Length (mm)	4,470	4,490	4,490	4,480	4,260	4,490	
Wheel-base (mm)	2,725	2,725	2,725	2,725	2,725	2,725	
Curb Weight (kg)	1,285–1,520	1,285–1,490	1,415–1,650	1,335–1,590	1,300–1,405	1,385–1,495	
Number Built	1,918,986	443,264	277,146	428,061	199,430	85,766	3,266,885

DESIGN UP CLOSE E46

After the big forward step represented by the E36 generation, the revisions for the follow-up E46 3-Series at first appear little more than a midcycle refresh. But as always with BMW's work, closer examination reveals the E46's design to be particularly well resolved—notably in its Coupe and Touring iterations. Attributed to Erik Goplen under the direction of newly arrived studio head Chris Bangle, the design language is exceptionally smooth, with barely a sharp radius in sight.

Wider tracks and wheel/tire combinations give the new design a secure, well-planted stance on the road, and on all but the lowest-powered versions, the wheels closely fill their arches to bring balance to the gentle wedge side profile.

The nose sees the hood wrap smoothly into the double-kidney grilles and faired-in lights, themselves with their four-lamp effect amplified by slim wave moldings separating them from the bumper. Prominence is given to the slightly raised grille and its body-color surround, and this was further strengthened on facelifted post-2001 models. The body sides, with their simple surfacing and straight-through feature lines, are spared any unnecessary detailing, all of which has allowed the shape to last extremely well and not to appear gimmicky or dated.

The interior, as so often with BMW, is a paragon of simplicity and good ergonomics, with discreet, high-quality materials, fine textures, and less of a wraparound "cockpit" feel to the driver environment than earlier generations.

Work on the E46 began after Chris Bangle's appointment as design director, yet the model's smooth, well-controlled contours show little sign of his influence. The clarity and simplicity were much appreciated in the market.

Wider tracks, wheels, and tires gave the new car a more secure stance on the road, and this was the model that introduced the stepped rear lights that would become a BMW brand signature for many years.

The two-door Coupe followed on quickly from the sedan. Lower, wider, and longer than its parent, it was again a completely fresh design by a different team. Notable is the near-perfect fit of the tires within the wheel arches.

was the expressed desire to shake up the "Russian doll approach" to design, where each model comes across as a slightly smaller version of the one immediately above it in the brand hierarchy. Technically, the E46 was the first BMW designed under his watch.

Yet in the end, Bangle must have chosen to take a back seat on the E46 design, for the work is attributed to another American, Erik Goplen, working at BMW affiliate Designworks in California. The new car was slightly longer and wider than its forebear, as well as more aerodynamically efficient; with all the added safety kit, despite the use of aluminum for the suspension, it emerged some 55 kilograms heavier model for model.

Yet whatever disappointment there might have been with the cautiousness of the design upgrade, it would be quickly outweighed by the major engineering advances made by this new generation. Most notably of all, the E46 was host to a trio of highly innovative new engines: the N42 four-cylinder gasoline Valvetronic, which brought new levels of performance, economy, and low emissions, and the M47 and M57 four- and six-cylinder direct-injection turbo diesels. These 2.0- and 3.0-liter units combined tremendous performance with astonishing economy in a way that had never been seen before, making diesels faster than their gasoline-fueled counterparts and totally revolutionizing the European market in the process.

But that's jumping ahead too fast, for the E46 had got off to a flying start even before these new engines appeared. A third of a million examples found customers in the 3's first full year on sale, with the Ci Coupe and Touring models now added to the mix. With three German factories running at full capacity, the annual tally soon began hovering around the four hundred thousand mark—unprecedented numbers for a BMW, and a reflection of the 3-Series' appeal across a broad range of segments. The car was especially popular in the corporate sector, where drivers of company cars prized its premium image and strong performance, and fleet managers were even more enthusiastic about its fuel economy and low lease costs on the strength of its high value retention.

But not everything was perfect, and right from the start there had been the familiar moan—in the specialist press, especially—that the 3-Series had "gone soft" with this model change and BMW had sold out to the luxury lobby. More stinging was the complaint that it simply didn't feel sporty like a good BMW should. And for once, there was a grain of truth in these assertions: The new model's suspension felt a good deal softer than before and, perhaps as a consequence, the steering seemed less precise. Of course, the opposite side of that coin was that the ride comfort and refinement were very much improved, and the serene interior encouraged comparisons with cars in the luxury class above.

NEW ARRIVALS

There were no great surprises with the Coupe and Touring derivatives, other than the wagon model arriving much more promptly than had been BMW's custom with past generations. The design itself was much as might be expected: a very similar overall look with just a minor stretch in dimensions, making it more practical than before but still prioritizing style over sheer storage volume. A useful innovation was the separately hinged tailgate glass, which allowed smaller items to be dropped into the cargo space without the need to open the whole tailgate.

The Coupe, by now carrying the Ci designation (which would last only one model cycle) was lower and wider than its sedan parent. As before, every panel and body component was different; the company boasted that the two-door shared only the door handles and the BMW badges with the four-door. The Coupe rode on wheels an inch bigger than the other models, the tires filling the arches with designer-perfect precision—something that helped highlight its elegant proportions.

A fresh Cabriolet soon followed, identical in format to the E36 version but with the added refinements of heated glass in the rear window, standard pop-up roll bars behind the rear passengers, and integrated seatbelts for the front seats to guarantee the optimum belt path—and to ease access to the rear seats. The Cabrio came with a hefty 10 percent price penalty over the equivalent sedan, along with a weight gain of 130 kilograms.

There then followed a bewildering sequence of engine and transmission changes, too many to recite here. These upgrades affected not only the basic power units but also their capacities and model designations. Eventually, the changes led to the groundbreaking Valvetronic and direct-injection diesel power units mentioned in the previous section. Also welcomed was a return of all-wheel drive to the 3-Series, which had been absent since the E30 generation in the 1980s. Dubbed xDrive, the new AWD system was taken from the X5 SUV, which was already proving a big hit in the market. It was simpler and more effective than the old setup and again proved popular in Alpine regions.

By this time BMW had rolled out its midcycle update—or Life Cycle Impulse (LCI)—across all but the Coupe, which already had a firmer chassis setup. This was a very light facelift, mainly affecting the grille, headlights, and turn signal lenses, which switched from amber to clear. More importantly, though, BMW had taken the feedback on board and turned its attention to the standard cars' suspension, dampers, and steering. This had a magical effect on the car's feel, making it firmer and more precise, and at no apparent cost to ride comfort.

(CONTINUED ON PAGE 100)

The interior of the E46 showed a marked step-up in quality, solidity, and simplicity, with the instruments now more closely grouped around the driver, and the advent of wheel-mounted buttons for audio, telephone, and cruise.

The whole cabin ambience was lifted to become more akin to that of a premium car from the class above, and the softer suspension settings translated into much-increased comfort for passengers.

This time round, BMW was much quicker to add the wagon version. The larger dimensions of the base model meant that this was the first 3-Series Touring large enough to qualify as proper family holiday transport; a very useful feature was the easy to open, separately hinged tailgate glass.

Second time around, BMW appeared to take the Compact variant more seriously. All versions featured identical dashboards and rear suspensions to the regular models, and the eager six-cylinder version counts as a hidden gem within the E46 catalog.

MOTORSPORT AND THE M3

High point: Many regard the E46-generation M3 as the finest of all. Beneath its typically understated exterior lay a 3.2-liter engine of simply breathtaking ability, allied to a chassis with just a smattering of electronics to keep an overzealous driver safe. What made this particular M3 such a success was its refinement and ease of driving in everyday traffic, morphing into superlative-melting performance and soundtrack when extended to its near-8,000 rpm outer limits.

How do you go one better when you've already been celebrated as the very best? That's the challenge BMW is constantly setting itself thanks to M Division's self-imposed rule that each new model it produces must outperform the one that it replaces. In the case of the third-generation M3, the challenge was especially difficult: The previous car had already made the big step of graduating from four cylinders to six, and from 2.3 liters to 3.0. What's more, the 3.2-liter Evolution version of that engine had already seen a further power rise, to 324 horsepower. But this being BMW, the engineers found a way.

The acclaimed 3.2-liter six was further tweaked up to 343 horsepower, giving it the highest specific power of any nonturbo production engine. The transmission was now a six-speed manual, with the option of SMG sequential operation. Again, as was usual BMW practice, apart from the trademark quadruple exhausts, the exterior giveaways remained subtle: lowered suspension, unique wheel designs, and discreet slotted air vents just aft of the front wheels. Underneath, the chassis received a full complement of M Division upgrades, many of them to near-race spec in view of the high speeds involved. Inside, the driver environment was comfortable and well specified, but still perfectly suited to the serious business of fast driving, on road and on track days.

And fast it was, staggeringly so. It was easy to drive gently, too, perfectly at home in town, but fast, open roads were where it most readily displayed its great talent, with an intoxicating rush of acceleration and a howl from the engine as it came on strong. The faster you drove, the more firmly glued to the road it felt—this was a car that just loved to perform. If there was one criticism of early examples, it was that the steering felt too light at low speeds, but this was soon attended to, and somehow it all gelled together magically.

If the "regular" M3 is more than the sum of its parts, then its evolution version, the CSL, is even more so. For, with the CSL, which revived an illustrious designation from the company's racing history, BMW's M Division achieved the impossible—it made an even better M3.

First off, a comprehensive race-style lightweighting program trimmed the car's mass by a remarkable 110 kilograms, and further work on the engine raised the rev limit to 8,200 rpm and power to 360 horsepower, at the same 7,900 rpm as before. Extra power was liberated through freeing up the induction system, which was now directly fed from the circular inlet found on the left of the air dam only on CSLs—one of the few identification points.

The only fly in the ointment was a surprising one: BMW's insistence on the sequential six-speed SMG II transmission, which hadn't gone down well on earlier models. CSL owners say that the SMG is something that takes acclimatization, and that it's good once you learn its ways.

BMW's work skimmed several tenths off the 0–100 km/h time, and top speed, in unrestricted form, leapt to 280 km/h. And if further evidence was needed, the CSL knocked half a minute off the regular car's already rapid Nürburgring Nordschleife time of 8 minutes, 22 seconds.

All this came at a price, of course, and there were gasps when the €85,000 tag was announced. Though fewer than 1,400 were built—compared to more than 85,000 standard M3s—the CSL is something of a halo car for BMW, as agile and hardcore racer-like as the original E30 M3, yet with greater speed and sophistication. It is lauded by many as the best M car ever, which is certainly saying something. Hardly surprising, then, that two decades down the line, quality CSLs are auctioning for almost double what they cost back in the day.

Just as with the E36 generation, the E46 version enjoyed a split competition career. The 1993 rule change for the prestigious European Touring Car Championship still held, effectively barring the 3.2-liter M3 from running, so BMW and its customer teams ran the 320i—with considerable success, taking several national and continental championships, including the European titles in 2003 and 2004. Works driver Andy Priaulx took the first three World Touring Car Championship titles in 2005, 2006, and 2007 in his 320i, and the M3 GTR scored one-two results in the Nürburgring 24 Hours in both 2004 and, with Priaulx at the wheel, in 2005.

The M3 competed in the GT class in the American Le Mans Series (AMLS) in 2000 but found itself outpowered by the Porsche 911s, as the technical air restrictor regulations had a much greater impact on BMW's engine than Porsche's larger 3.6-liter flat six. BMW's response was drastic: The Motorsport division developed a 4.0-liter V-8 to insert in place of the straight six. This gave the wild-looking M3 GTR much better torque and top-end power, and the Porsches were trounced. However, other teams protested the BMWs, and for the following season the rules were changed to require a much larger quantity of equivalent road cars to be built before the race car could regain its homologation. Regarding this as unfeasible, BMW withdrew, but the lessons learned on track with the M3 GTR were instrumental in making the road-going M3 CSL one of the finest sports coupes of all time.

A few key differences mark this interior as that of an M3, notably the revised instrumentation and, in this example, the optional SMG semiautomatic transmission. The robotized system wasn't an immediate success and required some acclimatization but, overall, this E46 generation M3 is prized as the last one before electronic systems began to take away from the mechanical purity of the vehicle.

Heart of the matter: BMW M engineers worked even more magic on the S54 B32 straight six to extract 343 hp at 7,900 rpm—and then to go one better with 360 hp for CSL.

Crème de la crème: BMW M engineers achieved the impossible with the CSL—to go one better than the already brilliant standard M3. The light, stripped-out, razor-sharp handling CSL got back to the original M3's racy roots and became a legend in the process.

(CONTINUED FROM PAGE 96)

It was still some way from qualifying as sporty, but that special niche was soon filled by the much-anticipated third-generation M3 (see sidebar on page 98). Launched initially only as a Coupe, the new M3 took the successful E36 formula and, implausibly, managed to stretch it further. With the acclaimed Motorsport-developed straight six now boasting a full 343 horsepower, the M3's performance was truly breathtaking—and so, too, was the customer response, despite its six-figure German price tag, roughly twice the cost of a regular six-cylinder model.

A FINAL SURPRISE GIVES E46 THE LAST LAUGH

When BMW finally divested itself of the Rover burden in spring 2000, everyone breathed a sigh of relief—not least the surviving board members. Bernd Pischetsrieder had been ousted as CEO the year before, and a safe pair of hands was at the controls in the shape of academic Joachim Milberg. The now-solo company could at last safely plan its own future undistracted by headwinds from the UK.

And as if to celebrate its newfound independence, BMW sprang a surprise on the industry with the launch of an all-new Compact, based this time on the E46. Until that point, the view in business circles had been that the previous Compact was at best a stopgap, and an uncharacteristically poorly executed one at that—though running changes had gradually made it more acceptable. But now, those pundits were forced to think again, and a certain logic to the new launch emerged. In the absence of Rover, this was again a defensive move, protecting the lower echelons of the brand until the BMW-inspired Mini arrived in 2001. It also anticipated the scheduled 2004 debut of the small "proper" BMW—the future 1-Series.

And just to make doubly certain that this new Compact was safe from being mocked for being out of date like the last one, the E46 rendition featured the full-on Z-axle rear suspension of the regular models (and the M3), and its interior was not a throwback to older editions, either. The final icing on the cake was that BMW chose the new Compact to be the launch model for its world-exclusive innovation in gasoline engines—the Valvetronic unit. This came as further proof of the company's seriousness.

It worked. The Compact II, while not much to look at, was small, youthful, and fun, as was its mission to younger customers. The new Valvetronic gasoline models were lively and agile, the 320td was pleasantly punchy, and the six-cylinder 323ti, with 192 horsepower motivating just 1,400 kilograms, was an absolute cracker. However, fewer than 200,000 of these Compacts were sold during their short spell on the market. The derivative may not have recouped its development costs, but the outlay was probably worthwhile in terms of recovered pride—and the reassurance that BMW wasn't being run by bean counters.

SAVING THE BEST FOR LAST

Almost without exception, the entire E46 lineup was on a high by the time it came for the sedan to leave the stage in 2004. The star performers, propelled by their innovative engines, were the 320d (which alone took 570,000 sales) and the 330d (with more than 133,000). The Valvetronic would have figured, too, had it arrived earlier, but of the grand total of almost 3.3 million cars sold—yet another all-time record for the brand—the six-cylinder models accounted for a good 1.1 million, or more than one third. This could be seen as proof positive that two of BMW's core tenets were still intact and working better than ever before: (1) investing relentlessly in R&D to lead the market in engineering innovation, and

The M3 cabriolet is a highly desirable high-performance sports roadster, which happens also to drive with ease around town and seat four in comfort, warmth, and safety.

A typical M3 interior, and pure BMW: not overladen with gadgets, not too showy or luxurious—but just perfect for the serious business of enjoying fast driving and exemplary engineering.

A range of high-specification forged alloy wheels ensured M3 owners could differentiate their cars from those of fellow enthusiasts for the marque.

COLLECTABILITY

The consensus among many BMW fans and marque specialists is that the E46 represents peak BMW 3-Series: the last "analog" generation with solid, classical design, minimum electronic intervention, and relatively simple maintenance. The roomy Touring wagon is a long-lasting workhorse (and rapid with six-cylinder engines), the Cabriolet is the last of the proper ragtops, and the 330i and 330d Coupes are sensible alternatives to an M3.

Standout model: M3 CSL. As the model that's frequently judged to be the best M-car of all time, how can the CSL not be the clear favorite? For sure, as a limited edition of fewer than 1,400, the prices are stratospheric, and the driving experience is off-the-scale brilliant if your preference is for a hard-line racer. But the plus side is that regular E46 M3s, which look almost the same, seem better value, and their thrills are more accessible. But look out for clever fakes and check the trunk floor carefully.

Hidden gem: Compact 325ti. This time round, BMW made sure the Compact had proper Z-axle rear suspension and a decent dashboard, and all the gasoline ti versions are handy driving tools. Of these, the 325ti, with almost 200 silky six-cylinder horsepower, is the gem that sparkles the brightest—but watch out for the usual signs of sports derivatives thrashed without mercy.

Wild card: 320i Cabriolet. Hardly a wild card in this very highly regarded pack, but keep an eye out in the listings for a four-cylinder ragtop with one owner, low mileage, always garaged and serviced by the book. It's still stylish and could be the bargain of a lifetime if you strike lucky.

ABOVE LEFT: **Standout model:** No doubt about it: The M3 CSL is the standout car in the E46 portfolio, and perhaps the best M3 of all time. Rare, with fewer than 1,400 built, it is a hardcore racer in temperament, and if you can find one, you'll be bidding serious money against wealthy collectors.

ABOVE RIGHT: **Hidden gem:** Yes, Compact again, but this time it's a hidden gem. BMW's second crack at a hot hatch was a vast improvement, especially with the peachy six-cylinder engine and properly sorted suspension. It kept a low profile, unlike its fun factor.

LEFT: **Wild card:** The four-cylinder Cabriolets are hardly the wildest of wild cards, but they are stylish, beautifully turned out, and surprisingly enjoyable to drive. Better still, they're less likely to have been abused by unsympathetic past owners.

BMW 3-SERIES E46: POWER TRAINS AND PERFORMANCE

Model	316i	318i	318d	320i	323i	328i	320d	325i	330i	330d	M3/M3 CSL
Engine Code	N42	N42	M47	M54	M54	M54	M47	M54	M54	M57	S54B32/ B32HP
Configuration	Inline 4, DOHC 16V, Valvetronic	Inline 4, DOHC 16V, Valvetronic	Inline 4, DOHC 16V, turbo, inter-cooler	Inline 6, DOHC 24V	Inline 6, DOHC 24V, Double VANOS	Inline 6, DOHC 24V, Double VANOS	Inline 4, DOHC 16V, turbo, inter-cooler	Inline 6, DOHC 24V, Double VANOS	Inline 6, DOHC 24V, Double VANOS	Inline 6, DOHC 24V, turbo, inter-cooler	Inline 6, DOHC 24V, double VANOS
Capacity (cc)	1,796	1,995	1,995	2,171	2,494	2,793	1,995	2,494	2,979	2,926	3,245
Fueling	Fuel injection	Fuel injection	Diesel direct injection	Fuel injection	Fuel injection	Fuel injection	Diesel direct injection	Fuel injection	Fuel injection	Diesel direct injection	Fuel injection
Power @ rpm	115 @ 5,500	143 @ 6,000	115 @ 4,000	170 @ 6,100	170 @ 5,500	193 @ 5,500	150 @ 4,000	193 @ 5,500	231 @ 5,900	184 @ 4,000	343/360 @ 7,900
Torque @ rpm (N m)	175 @ 3,750	200 @ 3,250	280 @ 2,000	210 @ 3,500	245 @ 3,500	280 @ 3,500	330 @ 2,000	245 @ 3,500	300 @ 3,500	390 @ 1,750	370 @ 4,900
Transmissions	M5	M5	M5, 4AT	M5, 5AT	M5, 5AT	M5, 5AT	M5, 5AT	M5, 5AT	M5, 5AT	M5, 5AT	M6/SMGII
Max Speed (km/h)	206	218	205	226	231	240	216	240	250	227	250/280*
0–100 km/h	10.4	10.0	10.6	8.3	8.0	7.0	8.9	7.0	6.5	7.8	5.2/4.9
Comments	From 2001 with Valvetronic	From 2001 with Valvetronic	Figures for post-2001 models with common rail				Figures for post-2001 models with common rail		M6 trans-mission from 2003	Common rail	*Without speed limiter

(2) guarding the brand image jealously to allow high volumes of a very rich sales mix without in any way imperiling the premium cachet of the marque. It's a precarious balancing act that few other players succeed in pulling off—but as we will see in the next chapter, even BMW was capable of a momentary wobble and giving everyone a fright.

Meanwhile, the first E46 to be replaced, the four-door sedan, was already on the off-ramp when BMW sprang its final and most memorable surprise. The consensus among the sporting cognoscenti had been that sports cars didn't get much better than the phenomenal, all-conquering M3. But BMW, ever in pursuit of higher standards, was determined to give the E46 a final flourish and go one better still. So the M Division engineers came up with an even lighter and faster version, one that revived the near-sacred CSL epithet from the glory days of the 1970s, 3.0-liter race-bred coupes.

At first there was skepticism, for this new limited-run derivative was priced at nearly double the already high figure of the regular M3 while also being deprived of much of its comfort equipment in the name of saving weight, and the engine power had gone up only fractionally. It seemed a massive price to pay for exclusivity—just 1,383 were built. But the driving experience (see sidebar on page 98), according to the few people privileged enough to sample it, was on another planet altogether, once again showcasing what magic spells the M Division specialists could cast. And if any further proof was needed, the CSL sold out instantly, and twenty years down the line those few cars are now selling at auction for double their original price. A fitting finale for what some regard as the best of all 3-Series generations.

Just how advanced BMW engineering had become was clear when even this humble post-2001 318i could boast Valvetronic variable valve timing on both overhead camshafts of its sixteen-valve engine.

M V8

THE MOOD MUSIC CHANGES

Not a car, but an
engine? One simple
reason: The fifth-
generation E90, which
debuted in 2004, was

A good volume of history needed to be processed in the lead-up to launching the fifth-generation 3-Series—history that colored the early 2000s with a curious blend of anticipation and trepidation. Anticipation because this was a commercially critical model, by far BMW's biggest seller, with some 40 percent of group registrations and a sizable chunk of corporate revenue. And trepidation on account of one man: BMW's new design head, Chris Bangle.

Bangle, aided and abetted by designer Adrian van Hooydonk, had produced a series of concept cars that both intrigued and baffled the design community, but shocked media commentators. The first big production car under Bangle's watch, the imposing flagship E65 7-Series of 2001, had also prompted widespread media and public outrage. But the fuss soon died down, and the pretty Z4 roadster and calmer, more mainstream E60 5-Series that came after were much more positively received.

The first 1-Series, arriving in 2004, again stirred things up, with its unusual pinched, hollow flanks and exaggerated cab-rearward profile. But what, precisely, would Bangle do for his first new 3-Series? Could he afford to take such risks in the high-stakes environment of a big-volume, big-money core model so central to BMW's prosperity? Or would he play it safe with a crowd-pleasing design?

The answer, which came in the shape of the new E90 3-Series sedan at the 2005 Geneva show, proved to be something of an anticlimax. The new car—shaped by Joji Nagashima, whose previous credits included the E39 5-Series—was clearly bigger than its predecessor. Yet with its low front, arched roofline and raised rear deck, its overall profile was still familiar. Even so, it seemed to lack the usual BMW personality, suggesting an uneasy compromise between continuity and change. The new and much weaker interpretation of the twin-kidney BMW grille was also widely criticized and would soon be revised.

The bluff rear aspect came in for negative comment, too. Most hurtful was the feedback that it reminded onlookers of contemporary volume hatchback models such as the Nissan Primera, Toyota Carina, and Mitsubishi Carisma.

Mention should also be made of the new interior design. Once again, this ripped up the earlier template and redistributed the elements in a wider, more horizontal layout. Instead of the information zones being closely focused around the driver, the new arrangement saw two separate clusters giving information—one in front of the driver with the conventional analog speedometer and rev counter, and a second central one featuring a screen display for the iDrive (also new to this class) and entertainment and navigation systems. BMW was very much ahead of the game in all of this, but again there were those who bemoaned the loss of the brand's famed "cockpit" feel.

ENGINEERING: WITH BMW, THERE'S NO ARGUMENT

While opinionated magazine columnists and keyboard warriors always seem intent on giving car firms free advice on the finer points of aesthetic design, when it comes to BMW and engineering, there can be no argument whatsoever. The E90 was a case in point, and once the focus shifted to engines, suspensions, transmissions, and electronic systems, the new car straightaway shone as a big step forward.

The fifth-generation 3-Series was clearly larger but cut a very similar silhouette to its immediate predecessors. The big differences were at the front and rear—and in the cabin.

Not the kindest aspect of the E90 sedan, and the subject of some debate after the launch. A succession of minor refreshes gradually dialed back its severity.

The new four-door was several centimeters bigger in all directions than its outgoing equivalent and carried a lot more in the way of safety structures, electronic assist systems, and emission control equipment; it also had many more luxury fittings. Yet through judicious use of lightweight materials—for instance, aluminum in the front suspension and high-strength steel for the new five-link rear axle arrangement—BMW brought the new car in at little more than the weight of the old one. Those innovative materials even included magnesium for the steering wheel core and in some cylinder blocks, too.

As for the launch engines—ten in all, evenly split between gasoline and diesel—they were mostly familiar from earlier models but heavily upgraded. These engines soon took up the pioneering Efficient Dynamics portfolio of fuel-saving measures first announced on the 1-Series in 2007, which included features such as only engaging the alternator and air-conditioning compressor under braking, and cutting the engine when stationary. Deceptively simple in concept and implementation, these measures combined to cut emissions and save the driver 20 percent on fuel bills—all without affecting performance or driver enjoyment in the least. Easily applicable to all models, it was a typically BMW masterstroke and caught competitors off guard. So much so, in fact, that rival industry bosses were openly fuming at not having thought of these ideas first.

EFFICIENT DYNAMICS IN ACTION

This was a time of massive progress and key industry breakthroughs in diesel and gasoline fuel injection, exhaust aftertreatment systems, turbochargers, and the sophisticated electronic networking of the engine and transmission as a single engineering entity. As a result of this technology race, with BMW often leading the charge, 3-Series engine specifications rarely remained static for long as performance-enhancing innovations were quickly applied. The key 2.0-liter diesel, for instance, was the most popular powerplant choice in the turn-of-the-century diesel boom and underwent multiple upgrades as technology progressed, even during the relatively short production lifespan of the E90 series.

As a graphic illustration of the progress racked up, this core 2.0-liter started with standard common-rail injection and little more than 100 horsepower. But improving diesel pumps and injector nozzles allowed the big leap to much more efficient direct injection; variable-geometry turbochargers and intercoolers helped provide better boost more of the time; and ever-higher injector pressures, allied with multihole injector nozzles and piezoelectric precision dosing of fuel, gave more refined, more powerful, and cleaner combustion. The outcome, as the E90 sedan prepared to hand over to its successor in 2011, was a 2.0-liter diesel that

BMW 3-SERIES E90: TIMELINE

2004	Four-door sedan announced
2005	Show debut of sedan as 318i, 320i, 325i, 330i; Touring versions follow later the same year
2006	Coupe (E92) announced, with completely different body; Cabriolet version switches to rigid metal folding hardtop
2007	New M3 launched as Coupe, sedan, and Cabriolet, with high-revving 4-liter V-8 and 420 hp
2007	335i Coupe added, with twin-turbo 6 giving 306 hp; M3 gets 7-speed M-DCT transmission option
2008	Facelift for sedan and Touring gives stronger frontal appearance; AWD option added to some models; diesel engines uprated
2009	Improvements to all engines; introduction of 320d Efficient Dynamics edition, the most efficient BMW ever
2009	Limited edition of 135 lightweight M3 GTS Coupes
2010	Facelift to Coupe and Cabriolet with reprofiled grilles and rear lights
2011	New-generation F30 models announced to replace E90 sedans
2011	Special 450-hp Carbon Racing Technology runout version of M3 sedan, with 7-speed DCT

BMW 3-SERIES E90: STYLES AND VERSIONS

Body Style	4-Door Sedan	2-Door Coupe	2-Door Cabriolet	4-Door Touring	M3	Grand Total
Length (mm)	4,531	4,612	4,612	4,527	4,615	
Wheel-base (mm)	2,760	2,760	2,760	2,760	2,761	
Curb Weight (kg)	1,400–1,580	1,435–1,715	1,670-1,905	1,505–1,750	1,655	
Number Built	1,878,728	380,695	254,050	588,861	65,985	3,102,334

BMW 3-SERIES E90: BASICS AND ENGINEERING

Layout	Front engine, rear- or all-wheel drive
Suspension F/R	Double-joint MacPherson strut axle/five-link rear suspension
Steering	Rack and pinion, power
Braking	Disc

DESIGN UP CLOSE E90-93

Following on from the well-received and highly regarded E46 generation, itself very clearly an evolution of the preceding model, expectations were that the 2005 E90 would mark a big step forward—especially as the more audacious Chris Bangle was now in charge of BMW design. Yet, when it arrived, it ranked as a rare disappointment for a BMW, smacking of a cautious compromise that had stepped back from the more radical ideas Bangle and his team had advanced on earlier models such as the 7-Series, Z4, and 1-Series.

Visually, the new 3 failed to convince. There seemed to be something missing in the now more fluid design language: Gone were the tension and the crispness of previous generations, and there were no obvious focal points to draw the eye. Key here was the new grille treatment, which appeared to relegate the twin kidneys to a secondary role in the model's frontal identity. The lower portions of the grilles were set farther back than the sweep of the hood coming forward, and their positions were only signaled by brightwork bars along their top edges.

The surface profiling of the car's body sides was also very new. Gone was the neatly disciplined cross-section of before, replaced by a more flowing interplay of convex and concave shapes designed to reflect the light in different ways. But again, there was no standout feature apart from the gentle longitudinal crease running through the door handles. This had distant echoes of the complex language of the Z4, as well as the much sharper pinch line on the recently launched 1-Series, but the effect was lost in most conditions, and the overall impression was one of anonymity.

It is significant that this generation of 3-Series was subject to more external facelifts than its predecessors. Most of these, such as the rethought grille treatment and the new engine hood with two pronounced ridges running rearward, were aimed at strengthening the model's personality and brand identity. The Coupe, for its part, was a more cautious but better resolved design, shaped by a different team and not subject to such far-reaching facelift action.

It was the nose of the new design that caused most of the fuss, with commentators and enthusiasts worried that the twin-kidney grilles looked weak and indistinctive.

The E90 sedan, with its complex body contours shaped under Chris Bangle's watch, soon revealed itself as a design unusually sensitive to body color and ambient light conditions.

Coupe versions of the E90 generation, now with their own E92 coding, were shaped by a parallel team and showed clearer style than the sedan.

boasted an unprecedented 180-plus horsepower and huge slabs of torque from very low rpm, all allied to remarkably low emissions and a frugal appetite for fuel.

As a final flourish, in late 2009, BMW readied a special Efficient Dynamics edition of the sedan equipped with a complete suite of efficiency measures that brought CO_2 emissions even further down to 109 grams per kilometer—quite an astonishing number for a largish car whose energetic performance was in no way blunted. The 109 figure helped place this edition in a very advantageous tax band for both private owners and company car users, further fueling the spiral of demand.

In 2010, BMW public relations was emboldened to place this technical progress on record, observing:

> It is interesting to reflect on just how far diesel technology has come in a dozen years. Back in 1997 the zenith of the diesel art was the BMW 325td Saloon, a car with 25 per cent more capacity than today's standard BMW 320d Saloon. The intervening years have made the 325td look sluggish and thirsty by comparison. Today's standard 320d Saloon, not [even] the 320d EfficientDynamics Saloon variant, has 58 per cent more power, 71 per cent more torque, 36 per cent lower emissions and consumes 58 per cent less diesel. At 7.5 seconds to 60mph it is also 4.5 seconds or 38 per cent quicker. That is progress.

Touring versions of the E90 appeared in the same calendar year as the sedans. The significantly roomier cargo area retained the useful separately opening tailgate glass.

The E91 Touring came at an important time when buyers were beginning to favor high-performance station wagons. This is the top 335d with a mighty twin-turbo common-rail diesel six rated at 286 hp.

PARALLEL PROGRESS: GASOLINE AND DIESEL

The pronounced market shift toward high-performing midsize diesels—at least in Europe—did not mean that development of gasoline engines and top-end variants in any way slowed down. During the currency of the E90 and its offshoots, the entry-level gasoline fours successfully combined direct gasoline injection (GDI) with lean-burn combustion to gain a clear 10 percent advantage in efficiency. Meanwhile, at the very pinnacle of the gasoline range stood the 3.0-liter 335i, which again broke new ground for BMW.

For one, it marked a complete break from the company's previous policy of not using turbochargers on gasoline engines. BMW's thinking had been that the poor pickup of turbos at low rpm would delay the response to throttle-pedal inputs and spoil the experience for the driver. However, the engineers' inventive solution earned the new engine the International Engine of the Year accolade several times in a row. Instead of one big and heavy turbo, which took a long time to spool up and deliver boost, they chose two much smaller and lighter turbochargers, each serving three of the six cylinders. These turbos came up to speed almost instantly, delivering full boost even at low rpm, thanks also to the pioneering Valvetronic fully variable valve control mechanism. The effect was almost instantaneous: The TwinPower Turbo engine stepped straight onto its massive plateau of torque from just 1,200 rpm all the way through to 5,000.

Not to be left out, the diesel range also gained a powerful flagship in the shape of the 335d—also twin turbo and based on the 3.0-liter, six-cylinder block. But this time the turbo arrangement was different: The two turbines worked in sequence, rather than in parallel, to provide consistent boost throughout the rev range. The outcome was impressive, with similar peak power to the gasoline 335i and huge reserves of torque at all speeds. The performance figures were broadly similar, too, underlining just how rapidly diesel technology had caught up with—and in some cases even overtaken—its long-running gasoline sparring partner.

OTHER SHAPES, OTHER NUMBERS

While all this engineering work had been going on in the background, more or less hidden from public view, the highly visible exterior profile of the 3-Series was beginning to fill out with additional shapes.

First, barely a year from the release of the sedan, there was the Touring wagon version. For the first time,

(CONTINUED ON PAGE 114)

TOP: Heads-up: The E93 Cabriolet was the first 3-Series to feature a multi-element rigid retractable hardtop, seen here midway through its cycle. Clever though the design was, many still objected. ABOVE: The main gripe was that while the car looked great with the roof retracted, with the roof in position, the car's proportions were no longer as pleasing as with a soft-top. And, worse, the complex mechanism was heavy, adding some 200 kg to the car's mass.

MOTORSPORT AND THE M3

BMW M Division engineers faced a tricky dilemma when planning the car that would replace the super-successful E46 M3. It would need more power in order to be on equal terms with its Audi and Mercedes-AMG rivals, the latter now beginning to install ever larger and more muscular V-8 engines—but the classic BMW six had been developed as far as it could possibly go, and it was unlikely to satisfy upcoming emissions regulations.

Two options presented themselves: the easy route to more power via turbocharging or moving to a bigger engine altogether. The latter was felt to be the most BMW-like solution, and the engineers turned to the M5's mighty 5.0-liter V-10 to derive a 4.0-liter V-8. Endowed with all the sophisticated control electronics of the V-10, the new engine was soon developing a comfortable 420 horsepower at 8,300 rpm, as well as more low-down torque than the six. What's more, it was 15 kilograms lighter and significantly shorter than the six, so it could sit farther back in the engine bay for better handling balance.

The considerable extra power prompted the M engineers to strengthen the E92 Coupe's structure, widen the tracks, and firm up the spring rates; some suspension bushings were replaced by steel ball joints and the steering rack was quickened. External differences included a larger power bulge in the already domed aluminum hood, a revised rear diffuser above the quad exhausts, and a carbon roof panel on Coupe versions. Initial models all had the six-speed manual transmission, but soon a seven-speed M-DCT dual clutch box became an option—the first time BMW had offered this type of gearbox. As with the bigger M5, there was a bewildering array of modes and programs for controlling engine and transmission response, dampers, and steering. In an instant, it seemed, the M3 had gone from the analog clarity of the E46 to the digital overload of a new electronic era.

In terms of the driving experience, the V-8 M3 provided very different sensations. The burbly engine was happy at low speeds but showed a strong preference for high revs, spinning well beyond 8,000 to deliver real energy. The M-DCT transmission was the perfect companion to keep it on the boil, and a thumbwheel allowed the driver to choose any style of gearshift, from auto smooth to racetrack harsh and all stages between. Yet as a driving experience, it was somewhat clinical, lacking the sense of involvement that came with previous generations.

And that, at the end of the day, is what would seal the fate of the fourth-generation version as an also-ran in the overall M3 popularity stakes. BMW tried very hard to push the model through special paint finishes and an unprecedented number of limited editions—eighteen at the last count—aimed at various national markets. Most were largely cosmetic exercises, but one, the GTS in 2010, is worthy of mention.

Much more extreme than the previous-generation CSL, the bright orange GTS was a stripped-out near racer, much lightened and with a factory-fitted full roll cage in place of the rear seats. The race-style rear wing and front aero vanes were adjustable. The engine was stroked out to 4.4 liters and 450 horsepower for a 3.6-second dash to 100 km/h—and all 135 quickly sold, despite the intimidating price.

All in all, some forty thousand M3 Coupes found buyers, with the Cabriolet and sedan versions bringing the total to almost sixty-six thousand—significantly fewer than either of its predecessors. On the racetracks, however, it was business as usual as the M3 made its comeback in 2012 to the German DTM series after a gap of twenty years. The season brought top results for BMW, the Schnitzer Motorsport Team, and Canadian driver Bruno Spengler: The BMW M3 DTM made a clean sweep of the manufacturers', team, and drivers' titles.

Already, there had been some good results in the American Le Mans series, where the 485-horsepower M3 competed in the GT2 class against machinery that included Ferraris and the dominant Porsches. Overall wins eluded the BMW team, now managed by Bobby Rahal's RLL setup, but there were many podiums and driver and team titles between 2010 and 2012, after which the Z4 sports car took over as BMW's track weapon. All this was good preparation for the return to European competition already mentioned, with class wins in the Spa 24 Hours, overall victory (BMW's nineteenth) at the Nürburgring event, and, of course, the now-famous M3 GT2 Le Mans race car painted by New York sculptor and artist Jeff Koons as the seventeenth BMW Art Car. It finished nineteenth overall in the 2010 event, hampered by its lack of an effective aero package for the ultra-high-speed Mulsanne Straight. The following year, the M3 GTE Pro took GT pole and twenty-ninth overall in qualifying and finished fifteenth overall and third in the GTE Pro class.

The E92 Coupe version arrived a year after the sedan, and this M3 a further year on. Both were entirely fresh designs, with the Coupe both longer, lower, and sleeker, and with bolder grilles.

Much prized by enthusiasts are the details exclusive to the M models, such as the side mirrors with two slender attachment arms, honed in the wind tunnel to optimize airflow and minimize wind noise.

The first, and only, V8 in a 3-Series powered the M3 to impressive acceleration and top speed with its keen 420 hp. But many felt the high-revving unit lacked the character of the straight six.

(CONTINUED FROM PAGE 111)
the derivatives were allocated their own type numbers, the Touring being labeled E91. Though it lacked memorable design cues—some cruelly likened it to a VW Golf station wagon—its load space clearly benefited from the larger footprint of the new generation, and the E91 sold well straightaway, becoming the second most successful of all the variants.

The Coupe took longer to arrive, perhaps for good reason. Assigned project code E92, it was a completely different design, based on the same platform but implemented by a different subset of the design team. The shape, penned by Marc Michael Markefka, differed quite considerably from that of the sedan. It was not only lower and sleeker but noticeably longer, too. Bolder grilles and lights helped the front look stronger; the body sides showed a more distinct horizontal feature line, dipping down just before the front wheel arches, and broader taillights emphasized the width of the rear. As before, not a single body panel was shared with the sedan or Touring.

Inside, the Coupe kept the same dashboard and controls as the sedan, but the seats and upholstery were upgraded. Mechanically, the suspension was firmer, the wheels were bigger, and the range of engines excluded the lower-powered choices—but for the first time on a Coupe, the options now included diesel units.

For BMW, the Coupes are always very important designs, as they form the basis for the whole marque's most important halo model—the high-performance M3. Much of the company's reputation among agenda-setting commentators and influential sporting enthusiasts revolves around the image and the performance of the M3 of the moment. For the E90 generation, that first came in 2007 with a glimpse of a show car concept, clearly very production ready but with little information about what lay under the engine hood.

The mystery was solved later that year when the real thing was revealed at the Frankfurt show. As with the E36 generation in 1992, this was a step-change moment for the M3—but now the graduation was from six- to eight-cylinder power. But this was no ordinary V-8, nor even one related to the 7-Series-derived V-8 fitted to E46-generation M3s for US racing. Instead, it was a 4.0-liter engine created by lopping two cylinders off the exotic Formula One–inspired 5.0-liter V-10 from the new M5.

The logic was impeccable. The V-8 was not only lighter and more powerful than the six it replaced, but it was more compact, too, allowing it to be placed farther back in the chassis for improved weight distribution.

The M3 also played host to BMW's first dual-clutch transmission, the M-DCT, which replaced the unloved sequential manual gearbox. Toward the end of the E90 series, this seven-speed unit also spread to the top end of the regular range, under the Sport Automatic label.

The new design themes for the E90 generation meant a change in dashboard architecture, too, toward a more horizontal layout. These were the first cars with the optional iDrive system, which meant a rotary controller on the center console and a center display module on the dash.

BMW did much to convince buyers that diesels could be just as sporting as gasoline-fueled cars, this twin-turbo six-cylinder motorization being a case in point.

With the Coupe already wider, longer, and arguably better resolved than the sedan, the M3 version added extra heft with wider wheel arches, a deeper front apron, and the usual M signifiers.

The dash of an M3 equipped with the M-DCT dual clutch transmission, the first on a BMW. The rocker switch behind the selector lever adjusts the sharpness of the shifts between the seven ratios.

The M-branded cooling air outlets on the sides of the M3—always a key distinguishing feature of the top performance model.

COLLECTABILITY

With the earliest E90 cars now twenty years old, many examples are on that tricky borderline between low-value commodity transport and potential classic. There are signs within the BMW community that this fifth-generation model could be coming into favor—so now could be a good opportunity to secure one at a reasonable price and in decent condition before it gets neglected—or abused—by further waves of cash-strapped owners.

Standout model: 335i. For once, it's not the M3, though the 335i's creamy direct-injection six, with over 300 horsepower, does come close. It's the most classically BMW of the range and, whether sedan, Coupe, or Touring, a fine car to drive. Its lower profile means lower prices, but for bigger spenders looking to play the market, any of the many M3 limited editions—the wilder, the better—could pay off in the long run.

Hidden gem: 318i. It is a tougher call for this recent generation, which never really caught the public imagination. But any entry-level four-cylinder will be an engaging, well-equipped, and sensible form of transport, all the while harboring future classic potential, especially if it's a Coupe.

Wild card: 320Si. This was a wild limited-edition 2.0-liter homologation special built to get BMW into the winners' circle of the World Touring Car Championship—which it did, thanks to star driver Andy Priaulx. The N45 engine, built at BMW's F1 facility, ditched the usual VANOS valve gear for a much higher 7,300 rpm rev limit and peak power of 173 horsepower on the road-going version. A six-speed gearbox linked to a short axle ratio gave eager performance at lower speeds—at least when the engine was working, which wasn't always. This car soon gained a reputation for simply eating engines, generally with cracked blocks. It's uncertain how many of the two thousand-odd cars built still exist, but any survivors that are fixable could make an interesting project for a brave enthusiast—not so much a wild card as a very dark horse.

Standout model: 335i: For once the standout car is not an M3 but something more discreet—in style if not in performance. Both 335i and 335d in whatever form have hugely energetic and enjoyable straight sixes and don't demand such intensive attention as the V8 M3.

Hidden gem: 318i: As so often with BMWs, it's the more everyday models that can provide unexpected driving enjoyment for modest cost. The 318i will do just that—and if it's a svelte coupe, so much the better.

Wild card: 320si (center): Not to be confused with the E30-era 320iS. This one is a crazy homologation special made to allow the likes of Andy Priaulx scoop the WTCC crown. Road cars are quick but terminally fragile, so surviving examples are at a premium.

Sedan versions of the M3 are sometimes seen as the poor relation to the more glamorous coupe but tend to be more affordable—and possibly less prone to abuse on track days.

ROUNDING OUT THE RANGE

The final piece in the E90 puzzle was the arrival of the E93 Cabriolet version in 2006, not long after the Coupe on which it was based. But this time the Cabrio was a very different animal, BMW having decided to abandon the familiar fabric soft top in favor of an engineered multipiece rigid folding roof system.

With the roof retracted, the new Cabriolet looked just as elegant and aspirational as the previous-generation model, even though the proportions of the trunk were not quite as comfortable. But with the roof raised—which could be done remotely via a button on the car key—some of the Cabrio's grace was sacrificed. The constraints of the rigid roof system forced a different relationship between the rear window and the rear deck that was less satisfying than that of the earlier soft-top models. Of course, the new roof arrangement further improved warmth, comfort, and sound levels inside the cabin, but it is significant that for the successor 4-Series Cabriolets, BMW would take the trouble to improve the roof-up proportions.

Another major but unavoidable drawback of the rigid folding roof system was its weight. Model for model, the E93 Cabriolets were almost 200 kilograms heavier than their Coupe equivalents, and this impacted their against-the-clock performance figures—something perceived as quite sensitive in top-end editions such as the M3 and 335i.

Close cousin rather than kid brother: The 1-Series, which launched in 2004 to replace the old Compact, used components pulled from the 3-Series—and won plaudits for its handling, if not its looks or its rear-seat space.

AHEAD OF THE GAME IN ELECTRONICS

It's a fair bet that there was not a huge takeup from the options list when the E90 range launched in 2005—especially from the top end of the list where the more exotic and less familiar equipment choices lay. The standard models were already well equipped, and BMW was determined to stick to its policy of "cascading down" innovative luxury and convenience features from its top-end models into its more accessible big-volume lines, such as the 3-Series. Today, convenience features such as one-touch central locking, satellite navigation, parking sensors, and information display screens are commonplace, even on cheap entry-level runabouts, but for the 3-Series in 2005 they counted as cutting-edge technical innovations.

First off, the 3-Series offered the pioneering iDrive equipment control system as an extra-cost item on lower-priced models, but standard toward the top of the range; this had first been launched on the luxury 7-Series in 2001 but later simplified in response to customer feedback. The 3-Series was the first in its sector with this type of system, but before long, most of its rivals began to bring their equivalents.

Another innovation at the time was built-in satellite navigation, linked to the new high-level display screen. Before long, this was enhanced with another first: onboard digital storage for satnav maps and the owner's own choice of music, uploadable via a USB port. Also offered were a head-up display, in-car internet connectivity, Bluetooth audio streaming, eCall, rearview camera, and live traffic and parking information. Safety systems, most of them standard, were a cut above the rest, too—as well as the familiar antilock brakes, dynamic stability control, and dynamic traction control, the 3-Series provided tire-pressure monitoring, cornering brake control, and hill-start assist. The adaptive headlights with cornering beams and high-beam assist are now also commonplace. This list was endless, but it demonstrated how BMW continued to push the boundaries of electronic and safety systems.

And on the ecological and air-quality side, too, the 3-Series stayed well ahead of rapidly tightening legislation on all continents. For example, when the 320d BluePerformance model debuted in 2008, it already fulfilled the much tougher Euro 6 exhaust standards set for 2014.

BMW 3-SERIES E90: POWER TRAINS AND PERFORMANCE

Model	318i	320i	325i	330i	335i	318d	320d	325d	330d	335d	M3
Engine Code	N46/N43	N46/N43	N52/N53	N52	N54	M47/N47	M47/N47	M57	M57	M57	S65 B40
Configuration	Inline 4, DOHC 16V	Inline 4, DOHC 16V	Inline 6, DOHC 24V	Inline 6, DOHC 24V	Inline 6, DOHC 24V	Inline 4, DOHC 16V	Inline 4, DOHC 16V	Inline 6, DOHC 24V	Inline 6, DOHC 24V	Inline 6, DOHC 24V	V8, 4ohc 32V
Capacity (cc)	1,995	1,995	2,497/ 2,996	2,996	2,996	1,995	1,995	2,993	2,993	2,993	3,999
Fueling	Fuel injection, direct from 2007	Fuel injection, direct from 2007	Fuel injection, direct from 2007	Fuel injection, direct from 2007	Direct injection, twin turbo, intercooler	Common rail VG turbo-diesel, inter-cooler	Common rail VG turbo-diesel, inter-cooler	Common rail VG turbo-diesel, inter-cooler	Common rail, VG turbo-diesel, inter-cooler	Common rail, twin VG turbo-diesel, inter-cooler	Fuel injection
Power @ rpm	127 @ 5,750 143 from 2007	150 @ 6,200 170 @ 6,700	218 @ 6,500 218 @ 6,100	258 @ 6,600 272 @ 6,700	306 @ 5,800	122 @ 4,000 143 @ 4,000	163 @ 4,000 177 @ 4,000 Later 184	197 @ 4,000 Later 204	231 @ 4,000 Later 245	286 @ 4,400	420 @ 8,300
Torque @ rpm (N m)	180 @ 3,250 190 from 2007	200 @ 3,600 210 @ 4,250	250 @ 2,750 270 @ 2,400	300 @ 2,500 320 @ 2,750	400 @ 1,300	280 @ 1,750 300 @ 1,750	340 @ 2,000 350 @ 1,750, Later 380	400 @ 1,300 Later 430	500 @ 1,750 Later 520	580 @ 1,750	400 @ 2,950
Transmissions	M6	M6, 6AT	M6, 6AT	M6, 6AT	M6, 6AT, 8AT	M6, 6AT	M6, 6AT	M6, 6AT	M6, 6AT	6AT	M6, 7DCT
Max Speed (km/h)	208 210 from 2007	220 228	245 350	250	250	206/210	225/230	235	250	250	250
0–100 km/h	10.0 9.1 from 2007	9.0, 8.2	7.0, 6.7	6.3, 6.1	5.6	10.6, 9.3	8.3, 7.9	7.4	6.7	6.2	4.8
Comments	All gasoline engines have Valvetronic from 2005.										

All in all, in spite of the relatively muted reception given to the earliest cars, this generation of the 3-Series really did the business for BMW. The outgoing E46 model established itself as the best-selling premium car in the world, and this performance was instrumental in boosting BMW to a highly symbolic milestone in its history: the moment in 2004 when it overtook archrival Mercedes-Benz to become the leading premium car brand worldwide. With production in three factories in Germany as well as in China, Mexico, Russia, and a host of other countries around the world, the high sustained sales pace of the E90 generation helped push annual BMW-brand sales past the million mark for the first time—another symbolic threshold.

When this generation began to leave the stage in 2011, with the sedans the first to be replaced by the new F30 series, more than three million had been built, a highly respectable result, even if it was somewhat short of the E46's lofty total. And even though, perhaps for the first time, there was no easily identifiable standout version—apart, maybe, from the zeitgeist-perfect 320d, which alone sold a million—the whole generation must be judged a quiet commercial success.

Yet the 3-Series story would never be quite the same again. During the E90's tenure, 3-Series sales had been holding steady at between two-fifths and one-third of the brand's overall tally, but the seeds of the 3's eventual decline had already been sown—by BMW itself. Those seeds would soon sprout into a new range of vehicles: the X5, X3, and later X1 SUVs, models that would soon ride a wave of worldwide popularity and begin to eat away at the market share of familiar genres such as sedans, station wagons, and even coupes. In addition, for the next generation, BMW would split the four-shape 3-Series range into two sub-nameplates: 3-Series, initially for the sedans and Tourings, and 4-Series, initially for the Coupes and Cabriolets. This, too, would clearly impact subsequent 3-Series sales stats, but, equally clearly, BMW didn't mind in the least. As long as customers stayed loyal to the BMW brand and not just a particular model, it was all profitable cash in the bank.

M GC 4420

THREE BECOMES FOUR—AND MORE

The great divide: The F30 generation saw the range split into two, with sedans and wagons under the 3-Series nameplate and Coupes and Cabriolets taking the new 4-Series badge. More variants soon appeared, such as the 3-Series GT and the 4-Series Gran Coupe, pictured.

This concept for a 4-Series Coupe (left) did the rounds of the auto shows the year before the real thing hit the showrooms and proved to be very close to the production model in overall feel. Noticeable changes for the volume build car (below) include a more humped hood, for pedestrian safety, headlamps extending into the grille surround, and a more emphatic treatment of the grille itself.

A decade into the new millennium, BMW was in fine shape. It had weathered the 2008–2009 financial crash rather better than its competitors, thanks to insightful production planning in anticipation of a possible market collapse that took others by surprise. The group was now reeling off quarter after quarter of rising figures, and CEO Norbert Reithofer insisted that the premium market would recover to precrisis levels. BMW's agenda, he said, had changed since the dark Rover days when managers hoped larger build volumes would bring the economies of scale necessary for the fat profit margins the stock markets seemed to be demanding. In the future, he said, the route to greater returns would be through higher volumes of high-priced premium cars of all sizes.

And the secret to that—and to Reithofer's ambitious group sales target of two million cars by 2020—was not only more factories and more markets (especially fast-growing China) but also a greater number of models to exploit the remaining niches in the premium market. The key player here would be the midsize range, still the group's bestseller overall, and still a major money spinner. But the 3-Series could not do the job alone. It needed to diversify its appeal to help it lay claim to every tiny corner of the market where premium customers might be lurking.

So when the next-generation 3-Series, the F30 sedan, was unwrapped in autumn 2011, BMW also announced a new naming strategy for its midmarket ranges. From now on, said the company, the 3-Series family would comprise sedans and Touring wagons, while coupes and convertibles—in other words, the more upscale models—would fall under the new 4-Series nameplate. Most agreed that this was a useful separation, as it would allow for greater product differentiation and a further climb upmarket in the future. Predictably, however, hardliners squealed that the psychologically all-important high-performance halo variant would no longer wear the hallowed M3 tag.

CHANGE ON ALL FRONTS

As befits BMW's engineering-led approach, the changes ran a lot deeper than the badge on the trunk lid. Not only was the new car much more striking and more expressive in its external style, but a lot had changed under the skin, too, and there would be plenty more advances in store as the years rolled past. For one, the fact that all the dozen-plus engines were turbocharged proved a strong headline grabber. For another, the standard gearboxes were six-speed manuals and eight-speed automatics across the board.

BMW 3-SERIES F30, F31, F34 AND 4-SERIES F32, F33, F36: TIMELINE

2011	3-Series sedan (F30) announced; all engines are turbocharged, TwinPower turbo gasoline, all 4-cylinder except 335i; transmissions are 6-speed manual or 8-speed automatic
2012	316i, 316d, 318d, and 320i versions added; F31 Touring joins range; xDrive four-wheel-drive and ActiveHybrid 3, with 6-cylinder engine and electric motor combination, become available
2013	F34 Gran Turismo launched with longer wheelbase, taller roof, fastback rear; five engine options from 143 to 205 hp
2013	F32 Coupe launched, now labeled 4-Series; F33 Cabriolet added, with revised three-section folding rigid hardtop.
2014	M4 Coupe launched, with 3.0-liter turbo engine and 431 hp
2014	4-Series Gran Coupe added; extended roofline allows five-door hatchback layout within same footprint as Coupe; M4 Cabriolet added
2015	Minor exterior facelift; new modular engine range now includes 3-cylinder gasoline; 330e plug-in hybrid replaces ActiveHybrid 3
2016	3-Series 325d gets new 4-cylinder diesel; limited-edition M4 GTS launched, with extreme race-style chassis and water-injection engine giving nearly 500 hp
2019	New G20 3-Series replaces F30 sedan

BMW 3-SERIES F30 AND 4-SERIES F32: BASICS AND ENGINEERING

Layout	Front engine, rear- or four-wheel drive
Power Train Types	Gasoline, diesel, hybrid, plug-in hybrid
Suspension F/R	Aluminum MacPherson struts/five-link rear axle in lightweight steel
Steering	Rack and pinion, electric assistance
Braking	Disc

BMW 3-SERIES F30 AND 4-SERIES F32: STYLES AND VERSIONS

Body Style	3-Series 4-door Sedan (F30), M3 (F80), and China LWB Sedan (F35)	4-Series 2-Door Coupe (F32) and M4 (F82)	4-Series 2-Door Cabrio (F33) and M4 (F83)	3-Series 5-Door Touring (F31)	3-Series 5-Door Gran Turismo (F34)	4-Series 5-Door Gran Coupe (F36)	Grand Total
Length (mm)	4,633	4,638	4,638	4,633	4,824	4,638	
Wheelbase (mm)	2,810	2,810	2,810	2,810	2,920	2,810	
Curb Weight (kg)	1,400–1,735	1,545–1,700	1,725–1,925	1,470–1,575	1,580–1,820	1,570–1,760	
Number Built	2,196,935	268,148	172,414	511,478	310,230	327,048	3,786,253

The sixth-generation 3-Series moved to a more complex, more expressive design language, but the overall impression given is still very characteristic. This is the M3 version, with deeper frontal openings, wider wheel arches, and side air vents.

Already, the whole lineup incorporated the full suite of Efficient Dynamics measures first seen on the 1-Series and the outgoing 3-Series. Now, in addition, all models adopted energy-saving electric power steering, all engines were future-proofed against upcoming emissions standards, and the pioneering ActiveHybrid 3 even joined the lineup.

And as if to demonstrate the platform's versatility in enabling future upgrades, over the course of its eight-year production lifespan, the model would host the first-ever three-cylinder engine in a 3-Series; the first modular family of engines; the first plug-in hybrid, the 330e; the first turbocharged M models in the midsize class; the first long-wheelbase sedans and family hatchbacks; and, finally, the first Gran Coupe sporting hatchbacks.

A RETURN TO STYLE

First things first: The four-door F30 sedan created a strongly positive reaction on its debut. This, ran the consensus, was an impressive return to form for BMW design, and there were no regrets at the passing of the previous generation. "We can relax now—the BMW we know and love is back," volunteered one observer.

Indeed, despite the new model's freshly low and wide frontal treatment, there was a comforting familiarity about the new 3's appearance, perhaps owing to the return of the forward-slanted shark nose and the clear, confident twin kidney grilles. The headlights, too, were drawn inward to meet the broadened grilles, in an echo of the full-width grille and lights combinations of the first and second generations. In addition, a strong "Zicke" pinch line, gradually strengthening toward the rear, gave real substance to the side view and helped disguise the car's increased size.

In round figures, the new car was just 100 millimeters longer than its predecessor, to the benefit of rear passenger space. But on the road, it gave a bigger impression still—almost to the point that it was hard to distinguish at a glance from the next-class-up 5-Series, which had been remodeled just months earlier. This appeared to fly in the face of Chris Bangle's mid-'90s assertion that he wanted to move away from the "Russian doll" approach to design, but no one seemed to mind. There were also some gripes that BMW's calling card of six-cylinder engines was represented by only a single model in the launch lineup, the 300-plus-horsepower 335i. The rest of the gasoline cars were all four-cylinder, even the 330i. But again, that didn't seem to put anyone off, and 3-Series sales took off rapidly: The car hit half a million sales in its first year, another record debut. Early press reports celebrated a return to form for the brand, with the UK's *Car* magazine declaring that the 3 had moved the game ahead and gone "straight to the top of the class."

"This is a really well balanced car," reported the journal, "one that revels in its 50:50 weight distribution and rear-drive layout, and one that feels so much lither than Audi's often leaden A4. It's also one that rides so well—it's almost a surprise to discover that it's still fitted with run-flat tires. This is surely the best-driving car in its class."

DIVERSIFICATION LIKE NEVER BEFORE

There was praise, too, for the new interior, with its central color screen, elegant instrumentation, and paler colors, selectable from the palettes embraced by the new scheme of three different trim themes—Sport, Modern, and Luxury. With the four-door sedan now well established and in high demand, and the just-launched Touring version at last offering a load space roomy enough to match those of Mercedes and Audi, BMW began springing the first of several surprises in its strategy of broadening the midsize car's market spread.

The first move was one that initially baffled the industry. With the market flop of the big and ungainly 5-Series Gran Turismo still fresh in people's minds, BMW tried precisely the same exercise with the 3-Series, placing a taller, wagon-like body onto the lengthened platform of the Chinese-market long-wheelbase 3-Series sedan. The extra 110 millimeters between the axles certainly resulted in a capacious trunk and near-palatial accommodation for rear-seat passengers, but there was no attempt to provide even token third-row seats. The higher silhouette, while not as clunky as that of the 5GT, also took the edge off the 3's much-praised handling precision.

In the end, the Gran Turismo sold at only about half the rate of the Touring—perhaps because there was confusion with the 4-Series Gran Coupe, launched the following year, which was a much sleeker five-door fastback version of the 4-Series Coupe that had itself appeared in the interim. The Gran Coupe was a critical and commercial success: It effectively served as a hatchback alternative to the sedan, appealing to family and business buyers at the same time, and it looked sportier than the Touring thanks to its lower roofline aided by frameless side windows.

As for the Coupe, it had been previewed by a concept version at January 2013's Detroit Auto Show; the production model arrived in the autumn of that year, to general approval. BMW PR perhaps oversold it somewhat, proclaiming "a new chapter in the history of elegantly sporty, two-door BMWs . . . the new model embodies the ultimate in aesthetic appeal and driving pleasure in the premium mid-size segment."

It had indeed stayed pretty faithful to the concept preview model, attributed to Won Kyu Kang. And while there was not as much differentiation from the parent sedan as there had been in the previous generation, it did come across as elegant and expressive enough to serve as the basis for the upcoming M4 edition. The Cabriolet derivative followed in 2014, still with a rigid folding roof but now with a much less clunky roof-up profile.

(CONTINUED ON PAGE 128)

Clearly visible in these studio rear three-quarter shots is the complex interplay of the surfaces on the 3-Series' body sides, with reflected light revealing the noticeable swelling of the rear wheel arch.

The swollen arches are even more pronounced in the high-performance M3 version with its wider tracks, wheels, and tires. The effect is to add a strong propulsive force to the car.

DESIGN UP CLOSE F30–36

It was all change for this sixth generation of 3-Series, both under the skin and on the exterior shape. Christopher Weil's design for the F30 marked a welcome and significant break from the somewhat shapeless E90 series, which had received a lukewarm reception. One of Weil's priorities was to reunite the signature BMW kidney grilles with the headlamps, which had been a strong feature of the E21 and E30, where a full-width background grille had held them together visually.

Weil's design smoothly links the widened and flattened kidneys to the main headlights—also wide—via extra driving lamps. The engine hood, already powerfully arched, dips down noticeably toward the nose, which is much lower than before, perhaps to improve pedestrian safety.

Below the bumper, the central air intake has been split into a deep vent on either side, linked by a shallow channel. The whole effect is to make the car look low and wide, with an impressively secure and grippy stance on the road. Side-on, the drooping nose is more clearly evident, but plenty of plan shape means the grille and headlamp are visible from the side, too, helping reduce the perceived bulk of the car. Slim roof pillars and the return of the so-called Zicke feature line along the side, linking taillights, door handles, and headlamp corners, help stretch the profile.

The Coupe, for its part, stayed close to the sedan in its body language and detailing, but with a clearly lower and more purposeful stance. Of the two five-door hatchback versions, the lines of the taller Gran Turismo were skillfully handled to allow it to look good in isolation (and certainly less bulky than the much bigger 5-Series GT). However, alongside the sleek and well-resolved Coupe-derived Gran Coupe, it immediately looked more prosaic and less exciting.

LEFT AND OPPOSITE BOTTOM: The 3-Series GT was the most controversial member of the expanded family. The whole lineup shared near-identical detailing, such as lights and grilles, but the taller stature of the roomy, longer-wheelbase GT made it more of a challenge for designers to disguise its extra bulk, especially from the rear. The only real identifier from the front is the deeper grille.

OPPOSITE TOP: The big move in brand-identity terms with the F30 generation was to extend the headlamp clusters inward to meet the grille surround, but there remained some awkwardness in the humped profile of the hood when viewed from the side.

M HG 5012

M KJ 7320

TOP: Touring versions of the 3-Series had grown steadily in size and stature over the generations, making later models fine long-distance haulers for business executives and families.

ABOVE: The M4 Cabriolet, showing the differences to the standard car such as the more complex front apron, the power dome on the hood, and the swollen wheel arches. New was an improved retractable hardtop, lighter, and, when raised, more elegant.

(CONTINUED FROM PAGE 125)

A FACELIFT—AND THE BIG ENGINE UPGRADE

By now all six different styles of the 3- and 4-Series were midway through their planned careers, and most versions had been on sale for several years—not that the models' popularity was in any way waning. But BMW always builds an LCI, or Life Cycle Impulse, update into all its product planning timetables, no matter how well those models are performing, and summer 2015 was the midrange series' turn.

The LCI models were given a facelift so minor it was hard to spot. It was just mild tweaks to the headlights, adding "eyebrow" daytime running lights to the top edge and replacing the optional xenons with even brighter LEDs; widening the grille, while the now-broader rear lights became LED as standard; and the offer of new alloy wheels and, inside, new choices for upholstery materials and decor woods.

But as always with BMW, the most significant changes were to the drivetrain. All four gasoline and seven diesel engines now belonged to the same modular family, with the capacity of each cylinder set at the optimum value of 500cc. These new units, all twin turbocharged and with Efficient Dynamics built in, carried the B prefix in their designations, replacing the outgoing N-labeled units in all but certain six-cylinder models. Thus, the 318i became a 1.5-liter, three-cylinder still giving the same 136 horsepower as the old 1.6-liter four. The 320i to the 330i were all new 2.0-liter, four-cylinder with outputs ranging from 184 to 248 horsepower. In terms of designations, the 330i took over from the 328i of the same capacity, while replacing the six-cylinder 335i was the 340i with a brand-new all-aluminum 3.0-liter six, also from the modular family.

Diesels told a similar story, with 2.0-liter fours in the 316d, 318d, 320d, and 325d offering between 114 and 218 horsepower. The latter increased the following year to 224 when twin turbos were added.

One epoch-defining change was the replacement of the 3.0-liter ActiveHybrid 3—a mild hybrid with a nonexistent electric-only range—with the 330e plug-in hybrid, which coupled the 184-horsepower combustion engine with an 80-kilowatt electric motor to give both powerful performance and a useful zero-emission range of 30 kilometers (a very good figure for that time). In some markets, special Efficient Dynamics versions of both the 320d diesel and 320i gasoline joined the range.

At the same time, the existing xDrive all-wheel-drive option was extended to additional models, and the eight-speed automatic transmission, available across nearly all the engine choices, now incorporated a fuel-saving coasting mode and interlinking with the navigation system to anticipate upcoming road features and select the right moment to change gear. And if that wasn't already enough cutting-edge technology, the built in full-color head-up display joined other driver-aid systems such as collision warning, automatic city braking, lane departure warnings, and driver alertness monitor to boost safety. Finally, the mild chassis tweaks to firm up the front end were lauded by testers for reducing roll and sharpening the handling poise.

M3: CHANGE TAKE TWO

The previous iteration of the M3, with its high-revving V-8 engine, hadn't gone down so well with the notoriously picky M-aficionados. In particular, the purists missed the soulful howl of the highly tuned straight-six engine as it redlined through the gears. What was more, the new platform was said not to be wide enough for V engines, so BMW chose to go back to a straight six for the fifth-generation version—except this time it was turbocharged to deliver 431 horsepower at its 7,300-rpm peak. The transmission was again the much-praised seven-speed M-DCT dual-clutch system.

And there was also the change of nomenclature to adapt to. M3s were now only four-door sedans, which many felt was wrong, while the mechanically identical M4 version was available in either of two familiar formats—the Coupe or Cabriolet. There was never an M3 Touring, in contrast to the G80 generation that would follow, nor did BMW give the M treatment to the stylish 4-Series Gran Coupe.

Despite all M Division's efforts, and in spite of the brand separation, the new M3 and M4 never fully satisfied the fans—until a series of evolutions finally restored it to favor.

(CONTINUED ON PAGE 134)

Riding in style: The sumptuous interior of a top-line Cabriolet showing the seatbelts built into the front backrests to allow easier access to the rear seats

ABOVE AND RIGHT: True to its name, the 4-Series Gran Coupe (above), with its lower roofline and frameless side windows, came across as wider, sportier, and more appealing than the larger GT (right). The GT, which wasn't continued to the next generation, was built on the longer Chinese-market wheelbase, giving palatial accommodation in the rear seats. Seen in isolation, it still carried enough of BMW's sporting genes to sell more than 300,000 units.

After the sedan and its long-wheelbase Chinese-market derivative, the elegant Touring was the third most popular F30-generation 3-Series. BMW kept it on the shorter platform in order to guarantee the company's sporting flavor.

September 2015 and BMW's Munich plant completes its ten-millionth 3-Series. Forty years since the original launched, the home factory is the only one worldwide to have built all six generations. Today, with four more plants also building 3- and 4-Series, the total is well over twenty million, with more than half of them the sedan version.

MOTORSPORT AND THE M3/M4

While BMW's decision to go back to six-cylinder power for the now-split F80 generation of M3s and M4s was broadly welcomed by M-fans, the prospect of turbocharging was greeted with some apprehension. And then there was also the new name for the familiar coupe flagship to get accustomed to.

Those reservations aside, there was much to admire about the new M4. It looked really good, neither too extreme nor too tame, and it gave out all the right signals, too. With two single-scroll turbos that spooled up quickly, throttle response was far more eager than the turbo skeptics had feared, and with 80 kilograms less weight and a searing 431 horsepower at 7,300 rpm, it was a prodigious performer. It also had an uncanny amount of grip at the front end and seemed to do everything perfectly.

Everything except, many complained, stir the soul like the good old E46 did. At fault here was the sound—or lack of it, especially when pushing the car hard. And because the new turbo motor had so much torque at low rpm, drivers didn't have to work the engine so hard, rarely needing high rpm to make rapid progress. All this led to the principal gripe—M-drivers never seem satisfied—that the car simply wasn't involving enough. It was objectively better than its predecessors, but it wasn't more enjoyable.

All that would change in 2017 with the arrival of the M4 GTS. A fearsomely expensive track-focused limited edition, it was if anything too involving—to the point that it was as exhausting on the road as it was sensational on a circuit. "It is the fastest production BMW there's ever been,"

Spot the difference: The GT4 version of the M4 coupe launched in 2014. Aimed at customer racers, it featured carbon doors, hood, and roof panels, as well as a competition exhaust and a powerstick system to adapt to Balance of Power rule changes.

reported Gavin Green in *Car* magazine. “It is sharp, agile, turns heads and assails the senses rather than assuages them. But it’s far from an unqualified success.”

Extensive weight-saving measures, coupled with race-spec chassis upgrades and water injection to boost the engine beyond 500 horsepower, made it massively responsive and a ferocious track weapon. But, concluded Green, “while the normal M4 is too soft, the GTS version has swung too far the other way.”

On the real racetracks, however, the M4 got off to a flying start in 2014. BMW made a big fanfare of the model’s comeback to the prestigious DTM racing series after an absence—at least in terms of official factory representation—of twenty-five years. Not only was the M4 the official safety car for the series, but Marco Wittmann drove his M4 DTM to victory in the very first race, at Hockenheim. It was BMW’s sixtieth win in DTM racing, and Wittmann went on to become the youngest champion in DTM history—and to clinch the team prize, too. He was champion again two years later, prompting BMW to launch a DTM Champion Edition of the M4, which predated many of the performance upgrades on the later road-going CS and GTS.

GT4 champions: The Gabriele Piana/Michael Schrey M4 crosses the Oschersleben finish line to take the 2021 ADAC GT4 championship (top). The duo stormed to victory in the first four races of the season. BMW returned to the DTM in 2014, with Marco Wittmann winning first time out. Customer team M4s (below) in the DTM series were influential in promoting road-going M4s.

All-wheel drive formed an important part of BMW's offer to customers, and acceptance soon spread well beyond mountainous regions and the US snow belt.

(CONTINUED FROM PAGE 129)

GOING OUT ON A HIGH

One of the truest tests of a model's standing—and stamina—in the market is how it is perceived when the time comes for it to leave the lineup. If it is already looking tired and outstaying its welcome, that could be seen as a bad sign, but if it is running strongly and the announcement that a replacement is in the offing prompts a "Why so soon?" reaction of surprise, then that's proof of a mission successfully accomplished. And there is no question about it: The F30 generation of 3- and 4-Series models emphatically falls into the latter category, in contrast to previous overlong generations such as the E21, which didn't leave a day too early.

In total, this sixth generation racked up nearly 3.8 million sales, with the 3-badged models alone (sedans, Tourings, Gran Turismo) accounting for more than 2.5 million. Another major contribution came from the Chinese-market long-wheelbase four-door, at 551,000. The 4-Series versions (Coupe, Cabrio, and Gran Coupe) accounted for 774,000, a comparatively small proportion of the overall generation total. The question of whether the 4-Series in any way cannibalized the sales of the 3 is an irrelevant one: Yes, the 4 borrowed some sales of cars that could have been badged 3, but all those sales represented money in the bank for the company. For BMW, the broader F30 family was a job well done—again.

And in any case, much more serious impacts were already looming as consumer fashion began to propel annual sales of the X1 and X3 SUVs beyond the total racked up by the combined 3- and 4-Series. Today, the combined total for the 3 and 4 hovers below 25 percent of group sales, compared with percentages in the mid-40s when the solo 3-Series was at its peak. Yet even those figures are misleading. The 3- and 4- Series' slice might be shrinking, but as the BMW cake grows ever larger, the number of midsize cars is close to all-time records—and everyone is happy.

ABOVE: True to the promise of all M cars, the new twin-turbo M4 was faster and more powerful than its V8 predecessor. But despite its prodigious performance, many felt that it fell short on driver engagement.

RIGHT: The fifth-generation M3 and gen-1 M4 were the first to use turbocharging, but not all enthusiasts approved.

ABOVE: Ahead of its time: The 3-Series Gran Turismo was BMW's second foray into the spacious luxury hatchback, and its skillful design has lasted much better than the awkward looks of its larger 5 GT predecessor.

LEFT: Integration: Electronics became more seamlessly integrated into vehicle systems, with navigation, driveline and chassis settings, and phone all controlled via the iDrive and center display. Note the keyless starter button and engine stop/start function.

BMW 3-SERIES F30 AND 4-SERIES F32: POWER TRAINS AND PERFORMANCE

Model	316i	318i	320i	328i	330i	335i	340i	Active-Hybrid 3	330e	M3/M4
Engine Code	N13	B36	B48	N20	B48	N55	B58	N55	B48	S55B30
Config-uration	4-cyl	3-cyl	4-cyl	4-cyl	4-cyl	6-cyl	6-cyl	6-cyl	4-cyl, e-motor	6-cyl
Capacity (cc)	1,598	1,499	1,998	1,998	1,998	2,979	2,998	2,979	1,998	2,979
Fueling	Gasoline direct injection, twin-scroll turbo; 335i and M3/M4 are twin turbo									
Power @ rpm	136 @ 4,400	136 @ 4,400	184 @ 5,000	241 @ 5,000	248 @ 5,200	302 @ 5,800	326 @ 5,800	302 @ 5,800 + 40Kw e-motor	184 @ 5,000 + 80Kw e-motor	431 @ 5,500–7,300
Torque @ rpm (N m)	220 @ 1,350	220 @ 1,250	290 @ 1,250	350 @ 1,250	350 @ 1,450	400 @ 1,200	450 @ 1,350	450 @ 1,450	290 @ 1,450	550 @ 1,850
Trans-missions	M6	M6	M6, 8AT	M6, 8AT	M6, 8AT	M6, 6AT	M6, 8AT	8AT	8AT	M6, 7DCT
Max Speed (km/h)	210	210	235	250	250	250	250	250	225	250
0–100 km/h	8.6	8.9	7.2	6.1	5.9	5.5	5.2	5.3	6.3	4.3

Model	316d	318d	320d	325d	330d	335d
Engine Code	B47	B47	B47	B47	N57	N57
Config-uration	4-cyl	4-cyl	4-cyl	4-cyl	6-cyl	6-cyl
Capacity (cc)	1,995	1,995	1,995	1,995	2,993	2,993
Fueling	Diesel direct injection, common rail, VG turbo; 325d, 335d: twin turbo					
Power @ rpm	116 @ 4,000	150 @ 4,000	190 @ 4,000	218 @ 4,000 224 from 2016	258 @ 4,000	308 @ 4,000
Torque @ rpm (N m)	270 @ 1,250	320 @ 1,500	400 @ 1,750	450 @ 1,500	560 @ 1,500	630 @ 1,500
Trans-missions	M6	M6, 8AT	M6, 8AT	M6, 8AT	8AT	8AT, AWD
Max Speed (km/h)	205	215	235	245	250	250
0–100 km/h	10.7	8,5	7.3	6.8	5.6	4.8

Comments	
	Data relates to post-2015 models only, except ActiveHybrid 3. All gasoline engines have DOHC, 4 valves per cylinder, VVT and VVL, direct injection, twin-scroll turbo. All diesel engines have DOHC, 4 valves per cylinder, direct common rail injection, single VG Turbo (except twin-turbo 325d, 335d).

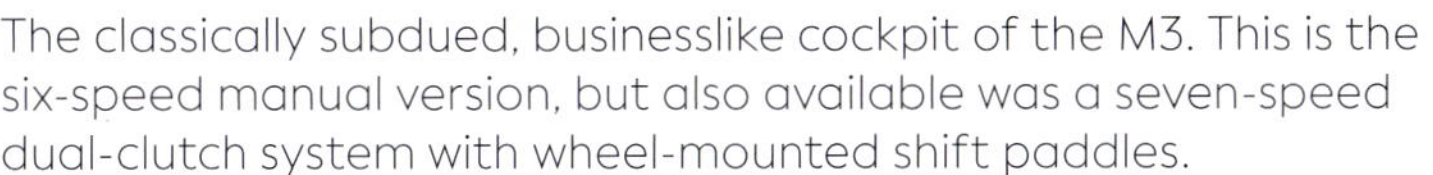

The classically subdued, businesslike cockpit of the M3. This is the six-speed manual version, but also available was a seven-speed dual-clutch system with wheel-mounted shift paddles.

This generation of 3- and 4-Series offered a record choice of sixteen different engines. Pictured is the top 3.0-liter 340i whose 326 hp propelled the sedan to 100 km/h in 5.2 seconds.

AT THE TOP OF ITS GAME

Return to form: The seventh-generation 3-Series was widely praised on its 2018 launch, but in a changing market it faced tough new opposition: BMW's own X3 and X4 sports utility models.

Bolder grilles were an immediate recognition point for the seventh-generation models. This one hides the new 374 hp 340i engine, one of only two in the launch lineup.

We now shift forward five years to late 2024, as this book is being written and, sixty months into its production run, the G20 3-Series is given a minor cosmetic refresh. This update provides a convenient moment to pause, take stock, and reflect on the whole 3 phenomenon so far.

At this juncture, the current model has already sold some three million units, comfortably bettering the sales pace of its F30 predecessor—itself a BMW sales record breaker with a total tally of 3.8 million if its 4-Series offshoot is also taken into account. Those twin performances, together with those of the preceding five generations stretching back to 1975, bring the overall lifetime sales of 3-Series models to well beyond twenty million. This further cements its place as the world's most-favored premium car and the segment leader in the three most significant markets globally: China, North America, and Europe. So the pivotal importance of the 3- and 4-Series to BMW is plain for all to see.

Yet it may not all be as straightforward as it seems. Looking at the bigger picture, the sudden rise to fashion of SUV-style vehicles means that the 3-Series is no longer automatically BMW's biggest-selling single model line; that position is under threat from own goals from the home team's X3/X4 sports activity vehicle combo. Even so, the shares of the 3- and 4-Series have been rising since 2021 to edge past the SUVs and regain pole position.

In any case, regardless of the ups and downs of the year-to-year statistics and the fast-fashion lure of the trendy SUVs, the 3-Series unquestionably stands out as the

Following in the wheel tracks of such a successful model, the new car had everything to prove—and critics' reports were universally positive, especially when it came to handling and strength of character.

emotional heart of the BMW brand, the car whose pioneering recipe and spectacular success propelled the company to the lofty position it enjoys today. In effect, it's BMW's equivalent of VW's Golf, and (is it really a coincidence?) both of these icons date back to first appearances that happened in the same year, 1974. Since then, the 3-Series has cultivated and then carefully nurtured the powerful brand values that now enable other BMW families, such as small cars and SUVs, to prosper in their turn. Beyond any shadow of a doubt, if there is a single most universally recognized BMW, it's a 3-Series.

THE WEIGHT OF RESPONSIBILITY

All this is to understand the daunting responsibilities that the BMW design team must have faced when, back in the mid-2010s, it began to scope out a new 3-Series for 2019—the seventh-generation edition of the corporate crown jewels. To add further momentousness to the task of upholding this legend, generation seven was the last one to be propelled by gasoline and diesel before the advent of the all-electric Neue Klasse in 2026, meaning that it would have to serve as a lasting monument to BMW's prowess in combustion-car engineering and design. That was no small challenge, and early rumors hinted at a radical approach establishing a new design language that would later migrate to other models.

BMW 3-SERIES G20 AND 4-SERIES G22: BASICS AND ENGINEERING

Layout	Front engine, rear-wheel drive, all-wheel drive
Power Train Types	Gasoline, diesel, mild hybrid, plug-in hybrid, fully electric
Suspension F/R	Double-joint MacPherson strut/five-link axle, coil springs
Steering	Rack and pinion, electric assistance
Braking	Disc

BMW 3- AND 4-SERIES G20–G83 : TIMELINE

2018	World premiere of G20-generation 3-Series at the Paris Motor Show
2018	M340i xDrive with new 374-hp engine unveiled at LA Auto Show
2019	Full range of G20 3-Series models announced at Geneva show: three diesel models, two 4-cylinder gasoline versions, and plug-in hybrid 330e
2019	M340i xDrive and 330e go on sale; Touring wagon models added
2019	Concept 4 study at Frankfurt show gives foretaste of upcoming 4-Series Coupe, including new grille style
2019	All-electric i4 version of 4-Series Gran Coupe previewed for 2021 production
2020	Mild hybrid assist added to 320d models, with 48-volt starter-generator and additional battery giving economy and performance boost
2020	4-Series Coupe (G22) launched; three gasoline and two diesel engines; 3-Series range gains 318i with new engine; 330e also available as Touring wagon
2020	4-Series Cabriolet (G23) unveiled, with new "bow soft-top" roof; three gasoline and two diesel engines; diesels and 6-cylinder gasoline M440i xDrive have 48-volt hybridization
2020	First details of new M3 sedan (G80) and M4 Coupe (G82) released; both have 6-cylinder turbo engine with 500+ hp, choice of automatic or manual transmission, rear- or all-wheel drive, 10-stage traction control
2021	All-wheel xDrive option added to 4-cylinder 4-Series Coupes; M440i available as rear-drive; new 3.0-liter diesel engine with two-stage turbocharging and 48-volt hybridization available across range
2021	Entry-level plug-in hybrid 320e models added, with system output of 204 hp
2021	4-Series Gran Coupe added, with engines from 420d to M440i xDrive
2022	Selected 3-Series models get Curved Display dash as part of a package of small upgrades
2023	4-Series models fitted with Curved Dash display, OS 8 and upgraded iDrive
2024	Updates for 3-Series models bring increased range for plug-in hybrid versions, Curved Display dashboard, OS8.5, and revised iDrive

BMW 3-SERIES G20 AND 4-SERIES G21, G22, G23, G26: STYLES AND VERSIONS

Body Style	3-Series 4-Door Sedan (G20)	4-Series 2-Door Coupe (G22)	4-Series 2-Door Convertible (G23)	3-Series 5-Door Touring (G21)	4-Series 5-Door Gran Coupe and i4 (G26)	Grand Total (through 2023)
Length (mm)	4,709	4,768	4,768	4,713	4,783	
Wheelbase (mm)	2,851	2,851	2,851	2,851	2,851	
Curb Weight (kg)	1525–1900	1,610–1930	1,770–2060	1,640–1,895	1,695–2290 (i4)	
Approx. Number Built to End 2023	770,000 +35,000 M3	110,000 +30,000 M4	65,000 +10,000 M4	270,000 +7500 M3	240,000	1,537,500

Shock value: BMW kicked the hornets' nest with the enlarged, thrusting grilles of the M4 coupe version, but few minded and most buyers reveled in the powerful image—if not the bulky rear flanks.

But when the finished product was unveiled at the 2018 Paris show, it showed clearly that those rumors had been somewhat exaggerated. This was not to be a revolution, but a sensible BMW-style evolution. The new G20 sedan, though longer and wider, immediately radiated a very strong 3-Series personality, with a confident, chrome-rimmed grille, a familiar 3-Series profile and proportions, and a secure, grippy stance on the road. Yet beneath those comforting design cues lay a lot of highly sophisticated attention to detail, especially in the surfacing of the sheet metal.

All this served to bolster the 3's stature as a smart and highly desirable luxury car, as indeed it was swiftly becoming: With its 83-millimeter jump in length, this car was now bigger than either the predecessor to the 7-Series in 1968 or the 5-Series in 1987. So, in every sense, it had a lot to live up to.

ENGINEERING: ELECTRONICS ON THE ADVANCE

As always, there were a lot more changes under the skin. The new car was built on the shared Cluster Architecture (CLAR) platform, which gave access to a fresh electronic architecture, too. This allowed the 3-Series to offer no fewer than forty driver-assistance aids, such as lane assist, parking aids, laser lights, and the new BMW Intelligent Personal Assistant, which could carry out voice commands and answer questions about the car itself.

Demand for diesel engines, especially in Europe, was already on the wane following the VW "dieselgate" emissions cheating scandal of 2015. Even so, three state-of-the-art diesel units still figured among the eight engines in the launch lineup, only two of which bore BMW's six-cylinder trademark. These modular engines, ranging from 184 to 374 horsepower, had already evolved to a high level during the life of the previous model, so updates were minimal. The standard transmission across the board was the eight-speed automatic, which delivered both better performance and better economy than other options. Only the two lowest-powered diesels offered the option of a six-speed manual, and even that would disappear in a subsequent update action.

Impressively, despite the 3's increase in stature and equipment, the new models weighed in some 55 kilograms lighter than their predecessors. This was down to the increased use of aluminum in the CLAR structure as well as in much of the suspension and running gear. Many of these parts, said BMW, were made with a large proportion of recycled raw materials, reinforcing the overall message of sustainability that BMW was beginning to develop.

(CONTINUED ON PAGE 146)

June 2020: The new 4-Series coupe is flanked by the classic 3.0CS and the 1930s 328 coupe at BMW's Munich studio. Design director Domagoj Dukec said the "striking vertical grille creates a bold and confident identity."

The smart Touring wagon once again proved a significant hit with customers and came second only to the core sedan in sales over the model's first four years on the market.

The intricate three-dimensional headlamps were typical of the effort BMW put into the detailed design of smaller components. Among the options at launch were laser lights and no fewer than forty driver-assistance aids.

DESIGN UP CLOSE G20-26

Coming on the heels of the outgoing F30 3-Series—which, with almost 3.8 million sold worldwide, had been the most successful model in BMW's history—the next-generation 3 had a tough act to follow. So it was perhaps surprising that early statements from BMW appeared to suggest a break with that successful formula, promising "radically revised proportions" and a new design language that would be gradually rolled out across the brand.

But the reality proved much less radical than implied, perhaps to the relief of the many who favored remaining with the core values of the outgoing car. The frontal impression of the new G20 series is if anything even stronger than before, with a broader and more confident chrome-rimmed grille extending smoothly outward to meet the lights. The lights are larger and more complex, and halfway across their width is a body-color upward ear in the front bumper to create a quad lamp effect.

The engine hood is powerfully convex, with strong raised contour lines converging forward onto the apexes of the grille; its surfacing is complex, too, with a slightly concave scoop between the contour lines. The body-side cross-sections are also very interesting, likewise featuring complex surfacing that begins at the front wheel arch and sees a taut ridge running rearward just below the glass line. Below this, a slight negative section runs through the door handles, below which is another raised line, itself crowning a further negative "light-catcher" surface before the rising line above the sill sweeps up to meet the rear wheel arch. A further detail is the swelling above the rear wheel arch, which starts at the door handle and runs back to the taillight. These sophisticated features show up best in strong light and paler body colors, but they demonstrate the meticulous attention given to this key design.

At the rear, the wider but slimmer taillights are still L-shaped and, thanks to darkened upper sections, help to disguise the height of the trunk lid and spoiler, the rear tapers noticeably in plan. Where the design is slightly awkward on some upscale versions is around the lower bumper and rear diffuser, where there's a lot going on.

OPPOSITE TOP: BMW had hinted in advance that the seventh-generation 3-Series would see a big shift in design language; in the event the proportions remained familiar, but new and sophisticated surfacing on the sides gave a fascinating interplay of reflections, light and shade, and a blend of subtle curves and sharp creases.

OPPOSITE BOTTOM: The rear view shows the faster rear screen and the elongated tail lights. Upscale versions lost some of this simplicity with a complex lower apron housing multiple vents, exhausts, and a diffuser.

LEFT: The big surprise promised by BMW turned out to be the Coupe, whose brash frontal treatment caused predictable shockwaves and contrasted sharply with the more fluid and voluminous architecture aft of the B-pillars.

M ZA 3023

M ZA 3023

(CONTINUED FROM PAGE 142)

POSITIVE REACTIONS

BMW has long been accustomed to getting broadly positive reviews each time it launches a new model, but this time round, there were almost no caveats at all. The more expressive design and detailing scored highly, as did the revised interior—even if there was some trepidation surrounding the Personal Assistant function. On the road, the combination of a stiffer structure and the clever new lift-control dampers brought a firmer but nevertheless very good ride on most surfaces. If there were any complaints, they tended to center on the steering, which, despite being adjustable for weight via the driver experience controls, could have done with more road feedback. And the rear seat still earned descriptions in the vein of "cozy" rather than "luxurious."

Writing in *Motor Sport* magazine, Andrew Frankel concluded that for a four-door sedan, the 320d was "improbably capable on a decent road. All cars like this grip hard and handle securely, but I doubt any of those you can buy new today are as deft and offer such steely body control. And while I'd prefer more steering feel, it's a car you can get into for the first time and drive quickly and confidently from the outset." Frankel ended his report by saying, "I welcome the fact that even this mid-range diesel is a 3-series of real character—BMW has put clear air between it and its rivals once more."

But BMW was not standing still, and a Touring version followed shortly afterward. At the same Frankfurt show in 2019, the company flagged up its sleek Concept 4 prototype, very clearly a blueprint for the forthcoming 4-Series Coupe spin-off. Interestingly, it had an early—and milder—version of the vertical grille that would cause such outrage when the real thing launched the following summer.

Meanwhile, new power plants, such as the 318i and 340d, joined the offerings. The latter replaced the 330d and had a new 3.0-liter six with twin variable-geometry turbochargers to deliver 340 horsepower and a massive 700 newton-meters of torque; unsurprisingly, it came only in M340d guise, with xDrive all-wheel drive as standard. A further significant refinement was the addition of 48-volt mild hybridization to the mass-selling 320d engine line, which not only improved economy and emissions but also allowed an extra 11-horsepower eBoost on hard acceleration. This win-win system would soon be extended to all engines in the lineup.

COUPE DESIGN SHOCKS THE ESTABLISHMENT

A showtime concept car traditionally presents a more extreme version of an upcoming production model, but when the first pictures of the production-spec 4-Series Coupe were released in summer 2020, the opposite proved to be the case. The outrage was instant, and it focused on just one feature: the grille. Between the Concept 4 and the

True to its role as a versatile carrier for all segments, the Touring was available with any of the range of modular power plants and transmissions—and even saw a rapid M3 version for the first time.

Elegance personified: The Cabriolet enhanced its status as a classy and rapid open-top four-seater, but by now it was becoming a large car—nearly half a meter longer than its E30 inspiration.

Concerns that the convertible 4-Series would look less stylish with its roof up were dispelled with this generation, which used a rigid so-called bow top to achieve a soft-top type profile.

A new era begins: The start of i4 production in 2021 gave BMW its first dedicated all-electric midrange vehicle. Though based on the combustion-fueled Gran Coupe version, it earned rave reviews.

The i4 topping up its batteries at an Ionity charging station. Along with other mainly European automakers, BMW is a stakeholder in this pan-European fast-charge network, located on major highways.

BMW has always stuck firmly to its policy of offering customers a choice of power trains in each model. Just like its Gran Coupe lookalike, the i4 came with the full suite of power and driveline options and M-style trim packs.

production car, the twin grilles had grown vertically, horizontally, and—critically—in perceived aggressiveness.

Now spanning almost the full height of the front, the sharp-cornered grilles were two gaping mouths hungry for air—a far cry from the discreet chrome-rimmed embellishment of the concept. This was a radical reworking of the brand's frontal identity, especially as a deep valley now trailed backward across the hood from the BMW emblem. Some worried that this "eyesore" grille would come to afflict the upcoming M3 and M4—which it did—and a German firm even began offering new sets of parts to reshape the model's nose.

But as is so often the case with these armchair-designer panics, the Coupe actually looked good in the flesh, if somewhat imposing, and the effect was less startling still in cars with darker body colors. Grilles aside, the rest of the Coupe was business as usual, with a 10-millimeter-lower ride height, firmer springs, quicker steering, and bigger wheels, and again the press reports were enthusiastic, especially about the M440i xDrive version. Many saw this as a very positive sign that the much-anticipated M4 would be a great one.

The Cabriolet (G23) followed shortly after, this time featuring what BMW bafflingly described as a new "panel bow soft top" that, the company claimed, combined the traditional feel of a fabric roof with the warmth, comfort, and convenience of a retractable hardtop. It was certainly less weighty than the previous solution—the Cabriolet now showed a much-reduced weight penalty over the Coupe of less than 50 kilograms. It looked good, too.

M3 AND M4: THE SAME, BUT DIFFERENT

These were, of course, the models that everyone had been waiting for. Each successive M3, and now each M4, was seen as a very public litmus test of BMW's expertise in its core mission of producing world-leading high-performance sports cars. The stakes were high, and constantly rising—after all, the previous M4 had topped out at 500 horsepower, and M Division's charter continued to insist that each new model outperform the last.

Yet, controversially, complaints that the outgoing model was short on driver involvement were addressed by offering a bewildering range of choices for the hapless buyer to navigate, not only the choice of standard or Competition engines, but also six-speed manual or eight-speed automatic transmission and, almost heretically, the option of driving all four wheels rather than just those at the rear. With each of these systems offering its own multitude of menus, sub-menus, and modes, the driver was confronted

(CONTINUED ON PAGE 154)

MOTORSPORT AND THE M3/M4

BMW needed to do more than simply go one better with this, the sixth-generation M3 and second-ever M4. Though the previous F80/82 model, the first to bring turbocharging to this size of M car, had been ferociously quick, critics and buyers felt it lacked the emotion and sense of driver involvement that makes for an authentic M model.

So it was surprising, to say the least, that in its initial information on the new models, BMW revealed that they would have not only turbocharged engines but also four-wheel drive and torque-converter automatic transmissions, the latter two features being widely regarded by the enthusiast community as anathema to the enjoyment of sports car driving. And then there was also the issue—extremely live at the time—of the new big-grille nose treatment. The prospects didn't look good.

Yet, true to form, M Division responded and managed to turn these developments to their advantage—and even to the approval of most enthusiasts. The requirement to be quicker than the outgoing car was settled swiftly with an enhanced version of the TwinPower Turbo setup, with each single-scroll turbo feeding a trio of cylinders. The cooling system now encompassed twin circuits for low and high temperatures, to the benefit of intercooler performance, and a divided oil pan took care of lubrication under sustained high-g cornering. Two engine tunings were offered: standard—as if it can be called that—at 480 horsepower and Competition at 510. Needless to say, the structure was reinforced in key areas and all chassis components significantly upgraded. Complementing all this was a host of driver-configurable systems affecting every conceivable area of the vehicle—engine and power train, chassis, brakes, aerodynamics, and comfort.

All semblance of subtlety went out of the window with the full-blooded CS version of the M3 sedan, especially in its more lurid color combinations. All-wheel drive was a controversial near-standard fit on all M3s, bringing weight to a hefty 1.7 tons.

In theory, a six-speed manual transmission and rear-only drive were also offered, but these were confined to the standard version, and most export markets chose only to offer the pricier Competition model—so in practice, very few customers were able to take advantage of those options. But once in the driving seat, all that was forgotten, and early reviewers raved about the blisteringly rapid acceleration and prodigious cornering grip.

As Andrew Frankel reported in *Motor Sport* magazine, "Even if you just set everything to 'comfort,' the (mechanically identical) M3 Competition is firm and no-nonsense. Structurally it feels solid as granite and offers that instant reassurance we crave from properly fast cars."

The new engine, he continued, is "obviously bursting with power, but what I like is how the additional urge has been extracted from the same capacity without dulling the throttle response or introducing any additional off-boost lethargy."

Yet even so, he conceded, at times he struggled to like the car, citing the lack of incisiveness of the transmission, especially on downshifts and even on its most aggressive settings. That was echoed by *Car* magazine, which loved the more linear power delivery of the new engine, with the spectacular top-end power taking over once the torque had fallen away. The magazine also declared the engine sound "much better" but, like Frankel, had concerns about the 1.7-ton weight and the eight-speed automatic, which it judged to be "not as confident" as the rest of the power train.

A strut—and engine bay—brace like no other: Raising the M3's power from 510 in the "standard" Competition version to 550 in CS form demanded extreme measures to ensure handling precision.

Even as late as 2022 BMW still had something fresh to pull out of the M3 hat—the first-ever production M3 Touring. Very powerful for its size, it was the ultimate high-speed load hauler. A CS version followed shortly afterward.

Pole position: Two M4 CS Edition VR46 coupes line up on the starting grid. Named in honor of the Italian motorcycle ace and now BMW works driver Valentino Rossi's favorite race number, the VR46s had special decor and came with an invitation to meet Rossi at his home.

Other reviewers were more fulsome in their praise, and it is interesting to note that very few of them ever mentioned the turbocharged nature of the engine, nor any notion that the four-wheel drive—deliberately set up with a strong rear bias—had in any way dulled the driving experience.

Mechanically upgraded special editions of this M3/M4 were limited to the promised "evolution" model, which in summer 2022 reintroduced the fabled CSL nameplate. Some saw this as a risky path, given how venerated the old 3.0 CSL coupe and later 2003 E46 M3 CSL had been. But the new CSL's credentials were impressive, with a 100-kilogram weight saving thanks to lots of carbon parts, a race-spec chassis, and a 39-horsepower power uplift. Impressive, too, was the price, at almost double that of the regular Competition model. But seen as a bonus was the deletion of the four-wheel drive and the fact that the edition was limited to 1,600 cars worldwide. And the claim that it had set the fastest ever Nürburgring lap time of any BMW production car.

Most reviewers judged it a qualified success, with hyper-responsive steering and a temperament more similar to the very hardline previous-generation M4 GTS than the E46 M3 CSL whose name it had borrowed. Yet, observed one, it took some years for the latter to become truly appreciated, so the M4 CSL could be a canny choice. An M3 CS followed in 2023, also adhering to the recipe of less weight and more power—in this case a full 550 horsepower from the uprated six. All-wheel drive was retained this time, as were four full-size seats, but the consensus among commentators was that this was not as successful an exercise as that applied earlier to the run-out version of the outgoing M5, also labeled CS.

The second-generation M4's motorsport career began in 2018 with the launch of the GT4 version, a €169,000 ready-made racer aimed principally at private customers and teams. As well as the usual carbon parts, such as the hood, roof, and doors, the GT4 had aero add-ons like front splitters and a rear wing. Unusually, it also pioneered power sticks, preprogrammed modules that would give different power levels—useful for race series where balance of performance rules applied. It had full race suspension, AP Racing brakes, Öhlins shocks, Makrolon screens and side glass, and built-in airjacks for quick pit stops.

Results came quickly, with the GT4 notching up numerous victories and titles in Europe, Asia, and North America in its first year, a winning streak that continues to this day. In 2019, BMW shifted its works support to the M6 GTS for long-distance racing while it worked on the design of the M4 GTS as well as a version for the brand's reentry into the high-profile DTM series. The M4 DTM began testing at Vallelunga in late 2021 and scored its first win at Mugello the following March; the GT3 won in America the next month, too. In 2023, both GT4 and GT3 took multiple wins on several continents, but while in Germany the DTM version powered to the 2023 drivers' title, BMW missed out on the manufacturers' championship. In 2024, following some early wins, the M4 began to struggle in the second half of the season. At the Le Mans 24 Hours in the same year, an M4 GT3 finished second in its class after a race-long battle for the class lead. And, incidentally, the much-hyped M Hybrid V-8s, one of which was the latest BMW Art Car by Julie Mehretu, were powered by an uprated turbocharged engine developed from the P66/1 2017–2018 M4 DTM motor. Competing in the top LMDh class, both M Hybrids qualified well but were involved in crashes early on; the Art Car was eventually repaired and managed to finish the race.

The M4 GT4 version proved so successful that in 2023 BMW commissioned an additional fifty to be built and announced plans for an Evo version. Lesser incarnations of the G20-generation 3- and 4-Series did well on the track, too, with the 325i winning the 2018 Nürburgring Endurance Series, the 330i M Sport taking the British Touring Car Championship six times in a row, and the 330e successfully taking over in 2022 when the rules allowed hybrids.

Rossi in action in the M4 GT3 EVO, qualifying for the 2025 Lusail World Endurance race in Qatar. The uprated 590 hp engine and driveline and aero tweaks for the new season helped the EVO to a win on its first outing.

The previous season's GT3 pauses in the pit lane. Among the upgrades brought in with the EVO version were smaller mirrors and a lighter cathodic dip rather than paint for the chassis.

(CONTINUED FROM PAGE 149)

with infinite permutations for how to configure the engine, transmission, suspension, steering, brakes, ten-stage traction control, and even the exhaust soundtrack for the journey ahead. And if the buyer had opted for the special M brakes, there was even the caliper color to choose, too—red, blue, or black.

In reality, it proved simpler than it first appeared. Most of the chassis settings could be memorized and recalled via the M buttons on the steering wheel, and the really big power train choices were often predetermined by the individual BMW importers of the buyer's country. In the UK, for instance, the only version available was the M3/M4 Competition with automatic transmission and all-wheel drive.

By this time, the shock of the grille had abated, and both the M3 and M4 received excited reviews. Pretty soon, BMW signaled that there would be a Touring version (G81) as well. No M3 Touring had been marketed before, though in the 1990s BMW had built a prototype E46 M3 Touring just for fun.

The curved dashboard display was a major innovation in information presentation, integrating instruments, navigation, and other information in a single slender, gently curved panel.

Current BMW models with the curved display still retain the much-loved console-mounted iDrive controller to operate key functions, but next-generation models may lose this feature.

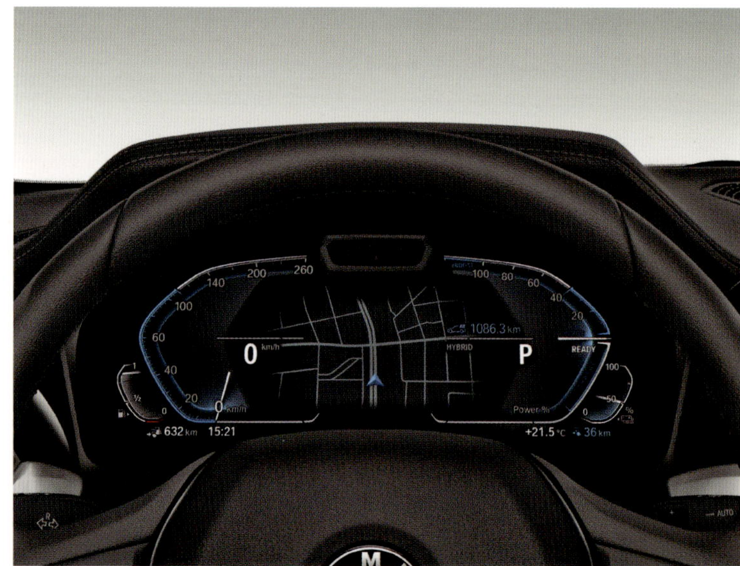

ABOVE AND RIGHT: The 3- and 4-Series moved to entirely solid-state instrument displays in this generation, allowing different views and selections. Not so immediately accepted was the anti-clockwise tachometer, but the central navigation screen worked flawlessly via the console-mounted iDrive controller.

Engineered with menace: The M4 (blue) and M3 (gray) in full attack mode. Even the most basic model boasted 480 hp, but most buyers opted for the faster 510 hp Competition or 550 hp CS versions.

THE i4 TAKES ON TESLA

As early as 2019, BMW had leaked select details of the i4, its all-electric rival to Tesla's Model 3. Finally launched in November 2021, it was effectively a battery-powered edition of the 4-Series Gran Coupe, which already existed in conventional combustion-engine form. Though crossing the weighbridge at well over 2 tons, it was extremely thoroughly engineered and quickly won plaudits from both the press and the business community. The M50 xDrive edition with twin motors and 544 horsepower was especially capable, storming to 100 km/h in 3.9 seconds—the same as an M3 Competition.

This caused a lot of combustion-engine die-hards to sit up and think. For many, the prospect of a future M4 running on electricity instead of gasoline suddenly became rather less alarming—see chapter 12.

All the while, the regular versions of the Gran Coupe had been doing what they do best: providing a more family-oriented alternative to the 3-Series with a sportier flavor than the Touring. Mention should also be made of the long-wheelbase Chinese-market version of the 3-Series, and in particular its battery-powered edition. Confusingly named i3, just like the German company's pioneering 2013 city car in Europe, the Chinese i3 was marketed in other territories such as India, too.

A round of updates to the 3- and 4-Series in 2022 and 2023, respectively, brought a revised interior featuring the BMW Curved Display, a highlight in larger models such as the iX. Linked to version 8 of the company's operating system, this provided enhanced voice and touch control to reduce the number of physical buttons on the dashboard—something that rarely goes down well, especially with older customers. But in this case, the complaints were short-lived

ABOVE AND RIGHT: The seventh-generation 3-Series was the first to have an entirely modular engine lineup, with six gasoline and four diesels from launch, all closely related. All ten were turbocharged and linked to standard eight-speed automatic transmissions, many also offering all-wheel-drive options.

and there was a guarded consensus that it actually worked well—in contrast to Volkswagen customers' nightmare experience with the ID.3 and Mark 8 Golf.

The facelift also served to increase the visual differentiation between the regular versions of the 3-Series and those with the M Sport option. With the latter, the frontal air intake grew much bigger and was pulled forward relative to the grille and lights for a more imposing look. A new steering wheel—two-spoked for regular models and three-spoked on M Sport editions—was also added. Important developments in battery technology brought extra zero-emissions range to the two plug-in hybrid models, the 320e and 330e; this took their zero-emission operating radius to 100 kilometers, crossing a key threshold that would allow them to perform most everyday work commutes without having to fire up their combustion engines. *Car* magazine declared the 330e to be "simply the best plug-in hybrid."

AN INTERIM RECKONING

At the time of writing, the G20 iteration of the 3- and 4-Series models still have a good few years to run before they are gradually phased out and replaced by all-electric Neue Klasse models. So it is much too soon to even attempt a definitive verdict on this, the seventh generation of BMW's sports sedan staple. But the signs certainly look good.

This 3-Series alone sold a solid three million cars in its first five years on the market. Add to this the growing contribution of the pricier 4-Series and of course the highly prized M3/M4 and it becomes clear that this iteration is on track to become the most successful of all seven generations so far. And most likely, in view of the richness of the overall model mix and BMW's super-slick manufacturing setup, the biggest money earner, too.

In technology terms, this has been the generation that honed the science of modular engineering down to a fine art, with substantial efficiencies in the supply chain as a result. It also brought astonishing combinations of performance, fuel economy, and clean emissions that no one would have dreamed possible just ten years previously. More critically, the G20 generation pioneered advanced electronic assistance systems that did not in any way dilute the all-important driver experience, and it seamlessly wove in widespread electrification—mild, plug-in, and full—to facilitate the eventual transition to battery power as an attractive option for a sports-minded clientele who might have been expected to scorn it. Yet above all else, true to the mission of the blue-and-white BMW roundel, this pivotal generation continues to champion the values of being fun to drive no matter which of the four fundamentally different power train choices the customer chooses.

But what lies around the next corner and the twists and turns beyond? How can BMW remain true to this lofty ethos in the upcoming years as battery power becomes the dominant motive force? And what does that imply for the eighth-generation 3-Series, due in 2026?

The broad-brush details are gradually coming into focus, and in chapter 12, we put together some pieces of the puzzle to provide a clearer picture of what to expect when the new generations roll out. But before that, for a full-disclosure background on what fires up BMW designers, engineers, and the global BMW community, turn first to chapter 10 for the inside story on the BMWs that failed to make it into the showrooms—and then dive deep into the BMW fan zone of clubs, collectors, fixers, and wheeler-dealers who live and breathe the brand every waking moment in chapter 11. Don't miss it.

BMW 3-SERIES G20–G26: POWER TRAINS AND PERFORMANCE

Model	318i	320i	330i	330e	M340i	M3/M4	318d	320d	330d	M340d	i4
Engine Code	B48	B48	B48	B48	B58	S58	B47	B47	B58	B57	Gen 5
Configuration	4-cyl	4-cyl	4-cyl	4-cyl	6-cyl	6-cyl	4-cyl	4-cyl	6-cyl	6-cyl	Electric single/twin motor*
Capacity (cc)	1,998	1,998	1,998	1,998	2,998	2,993	1,995	1,995	2,993	2,993	—
Fueling	Gasoline direct injection, twin-scroll turbo; M3/M4 have twin turbos						Diesel direct injection, common rail; 330d and M340d have twin sequential VG turbos				—
Power @ rpm	156 @ 4,500-6,500	184 @ 5,000-6,500	258 @ 5,000-6,500	184 @ 5,000 + 113 hp e-motor	374 @ 5,500-6,500	480 @ 6,250/ 510 @ 6,250*	150 @ 4,000	190 @ 4,000	265 @ 4,000	340 @ 4,400	340/544*
Torque @ rpm (N m)	250 @ 1,300	300 @ 1,350	400 @ 1,550	420 @ 1,350	500 @ 1,900	550 @ 2,650 650 @ 2,750*	320 @ 1,500	400 @ 1,750	580 @ 1,750	700 @ 1,750	430/586*
Transmissions	8AT	8AT	8AT	8AT	8AT	M6, 8AT	M6, 8AT	M6, 8AT	8AT	8AT	—
Max Speed (km/h)	223	235	250	230	250	250/290	218	235	250	250	190/225*
0–100 km/h	8.4	7.1	5.9	5.8	4.4	4.2/3.9*	8.3	6.9	5.5	4.6	5.7/3.9*
Notes		320e from 2021 has 204 hp system power				Max speed without/with Driver's Pack *Figures for Competition version					*Data for i4 eDrive40 except for i4 M50
Comments	All gasoline engines have DOHC, 4 valves per cylinder, fully variable valve timing and lift (double VANOS and Valvetronic). Diesels also have DOHC, 4 valves per cylinder.										

Parting shot: The 2024 M4 CS Valentino Rossi special edition. Fears that this would be the last-ever gasoline-fueled M4 proved unfounded when BMW announced that the successor-generation M3 and M4 would not be entirely battery powered but would include combustion-engined models too.

WHAT MIGHT HAVE BEEN

Face of the future: When Paul Bracq's other-worldly Turbo concept was unveiled ahead of the 1972 Olympic Games in BMW's home city of Munich, onlookers gasped in amazement. Though it would never be built, the striking design had a profound influence on BMW cars that followed—even the legendary M1 supercar.

From the revolutionary prewar 328 sports car to the glamorous 507 roadster of the '50s and the iconic first-generation M3 in 1986, countless fabulous designs mark out BMW's rich 110-year history. The world owes BMW a debt of gratitude for these and many other genre-defining models. But what of the unseen designs, the ambitious projects that never saw the light of day—perhaps because they were too radical or too expensive, or because they simply weren't aligned with the message the company wanted to project at that time?

Here, with a big shout-out to author and former BMW consultant Steve Saxty, we present a handful of the might-have-beens scooped up from the cutting-room floors of various BMW studios around the world. From the first early-'50s city car that could have been a bigger hit than the Fiat 500 or the Austin Mini to the hot 2002 remake torpedoed by troubles at Rover, we detail some of the gems that were denied the chance to sparkle in showrooms and on customers' driveways. Some were way-out wacky, while others fell victim to disagreements among top-level managers. Still more bear witness to the incredible excitement and design imagination that goes into anticipating the mindset of BMW customers ten or twenty years into the future.

And bear in mind, too, that in focusing purely on the designs that either are associated with the 3-Series or fit into its market segment, our selections represent just a tiny fraction of the BMWs that might have been. Scores of others—supercars, luxury sports cars, muscular off-roaders, innovative city cars—will have to wait for a later compilation of unseen and unreleased concepts from BMW Design's secret back catalog.

Thanks in no small measure to Saxty's never-before-granted access to BMW's Munich design studios and his extensive discussions with the com-

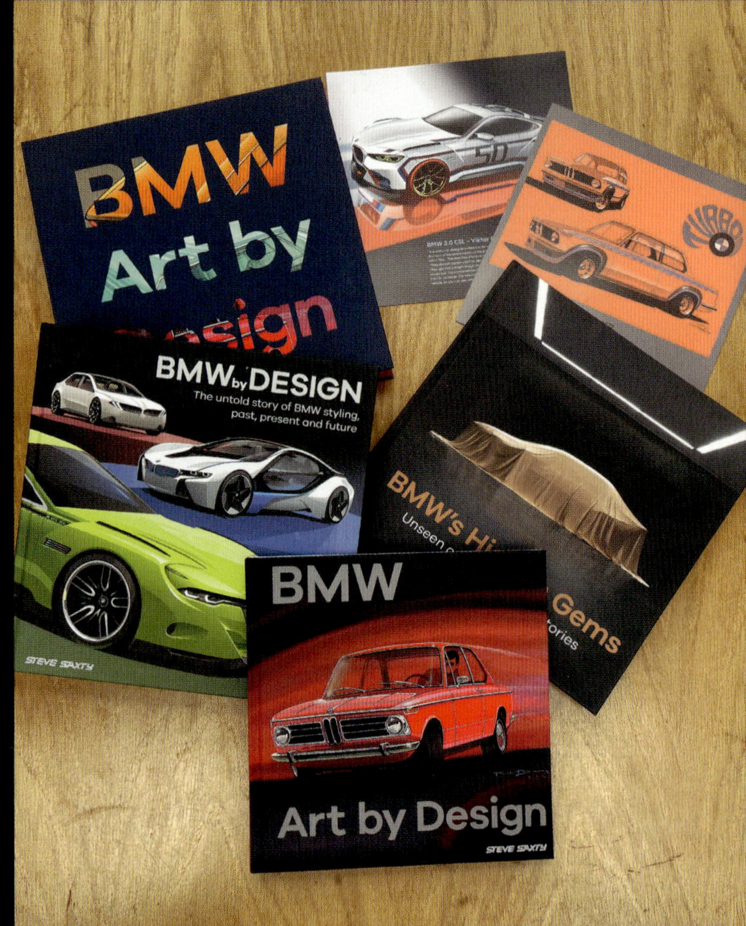

The three-book set *BMW Behind the Scenes*, by Steve Saxty, who was granted unprecedented access to the firm's design studios and documentation. Several of the sketches he was shown are featured in this chapter, with kind permission.

BMW's 1950 riposte to the all-conquering Fiat Topolino had the makings of a world beater, but company bosses canceled it in favor of returning to prestige cars promising bigger profits.

331

The 1950 baby-car project that could have been bigger than the Fiat 500 or Austin Mini

BMW was on the rocks in the late 1940s when this svelte 331 prototype was being developed. The firm was allowed to build only small-capacity motorcycles with thin profit margins, and in any case the public mood was shifting to small cars instead. Chief engineer Alfred Böning's sleek two-seater design made commercial sense, using the flat-twin motorcycle engine up front: It would have been cheap to build and, with the right marketing, could have been a bigger phenomenon than the 500 Topolino or even the Mini.

The idea made industrial sense, too. The design, with its standard rear-drive configuration and familiar engine, would have integrated easily with existing factory facilities. The twin-cylinder boxer motor was already proven in its reliability and, as was later shown to excellent effect in the 700 coupe, it offered great tuning potential. Finally, there was the undeniable fact that the European economy was at last beginning to recover, and the market was hungry for affordable small cars.

Nevertheless, BMW's board demurred, and it provoked something of a split when the company's top brass chose instead to build a big car that harked back to the firm's prewar glories. Their reasoning was that better profits were to be found in building high-priced luxury sedans. So, the 331 was scrapped and the lumbering, overbodied and underpowered 501 "Baroque Angel" got the nod, only to struggle in the marketplace and push BMW deeper into the red. Ironically, it was another pint-sized model—the 1955 Isetta bubble car—that stepped in to keep the marque afloat long enough to come up with the fresh designs that would finally put BMW back onto a path to prosperity.

Italian car couturier Bertone's late-'60s proposal for a 2002 replacement still looks fresh and modern today, apart from the exaggerated depth of its windows. The beginnings of the shark nose are clear and would inform the first 5-Series, which appeared in 1972.

GARMISCH

Italian flair that sparked the 3- and 5-Series phenomenon

Fast-forward to the mid-1960s, and BMW is on an astonishing—and somewhat unexpected—roll with its smart four-door 1500–2000 sports sedans and their smaller, even hotter-selling two-door '02 series offshoots.

The leading Italian design houses had been in on BMW's act from the start, with Michelotti penning the 1500 and Bertone shaping the elegant larger luxury coupes. Quite naturally, the Italians' thoughts now turned to what might eventually replace the '02 and the four-doors, and among the many proposals submitted confidentially to BMW was one from leading couture house Bertone. On the drawings was the signature of a certain Marcello Gandini.

When BMW failed to take the bait, Bertone began exhibiting the elegant and innovative prototype on the motor show circuit under the highly provocative label of Garmisch—which just happened to be a town close to BMW's headquarters.

Yet even though BMW apparently preferred its in-house designs, the Garmisch did end up having a profound influence on the company's style going forward. Though the Bertone study was purportedly for the smaller E21 3-Series project, many of its fresh themes showed through in the first 5-Series, which debuted in 1972.

Despite its clear significance, it emerged in 2018 that Bertone's prototype had disappeared without a trace. So important did BMW's current studio bosses believe it was to the company's design story that they commissioned Turin prototype builders Superstile to recreate the Garmisch using only Bertone's original drawings and a 2002 tii donor car provided by BMW Classic. The result, after many months of intensive work and painstaking parts sourcing, is a highly authentic and fully drivable car that, with its smooth lines, clean surfacing, and balanced proportions, looks surprisingly fresh and modern, even to today's eyes.

Indeed, the Garmisch's vertical rendition of the twin kidney grilles is referenced in one of BMW's very latest Vision studies—that for the crossover/SUV "X" version of the upcoming all-electric Neue Klasse generation. Who knows? This may become the signature feature of all future X-series BMW models.

Today's BMW designers were so impressed by the Garmisch that they built this fully drivable replica around a 2002 tii chassis. Only period-specific details such as the hex mesh over the rear screen betray its sixty-year age.

By the late 1960s BMW was toying with the idea of a mid-market roadster to take on the Brit sports cars. Independently, Bracq's vision, above, wasn't taken further—but it led to him being hired to lead BMW's design department.

E19 ROADSTER

A freelance designer's visionary sketch that got him the top job at BMW

In the late 1960s, BMW was toying with the idea of a small roadster that would be much more affordable than its aging heavyweight Coupes and Cabriolets. Spurred on by its perceptive US importer Max Hoffman (who a decade earlier had triggered the now-legendary 507 project) as well as energetic sales director Paul Hahnemann, the company contacted French design studio Brissonneau & Lotz to explore possible ideas. Part of B&L's allure had been the presence of French-born designer Paul Bracq, who had created the 230 SL "Pagoda" sports car while at Mercedes.

Although the B&L group was soon dismantled and nothing concrete came of the roadster project (which was given the code name E19), this and other sketches of the roadster proved to be of pivotal importance within BMW. Hahnemann quickly identified the sketches' author, Bracq, as being an exceptional talent. A full-scale model had been built up, complete with a highly innovative interior treatment; this was enough to persuade BMW's board to hire Bracq as the company's first-ever director of design. The novel driver-focused dashboard design soon found its first airing in the 1972 5-Series, to widespread press and public acclaim.

The roadster itself was never built, but, relates Steve Saxty, on B&L's breakup the project was moved to Bertone. Perhaps coincidentally, some of the design's themes later found echoes in Fiat's nimble X/19 sports car (also Bertone) and Alfa Romeo's Alfetta sedan series. Yet there is the lingering suspicion that had the BMW roadster been granted the green light, it could have given the fashionable Alfa Spyder and the aging British soft-top sports cars a seriously good run for their money.

Steve Saxty's book *BMW Behind the Scenes* reveals the illustrative flair of Paul Bracq. Here is another roadster proposal, crisp and modern in its body language, and with overtones of the Fiat X1/9.

Bracq's early work at the BMW included this sketch for the E21 project for the 2002 successor. It got the nod, but newly hired sales director Bob Lutz made Bracq swap the hatchback rear for a standard trunk. Spectacular sales proved Lutz right.

E21 3-SERIES HATCHBACK

Ahead of its time, but not a fit with BMW's 1970s master plan

It was in the mid-1960s that the hatchback body style first began to appear in the European market, sparked by the Renault R16, the R6, and, in 1972, the mass-selling R5. BMW was always well above these models in its pricing, but among Bertone's submissions for the early-1970s replacement for the '02 models (see the previous section on the Garmisch) were some proposals that included a hatch-style opening tailgate. Freshly arrived design director Paul Bracq was enthusiastic about the idea, and his sketches were enough to convince several BMW executives—until the arrival of Bob Lutz.

In his role as sales director, multilingual Swiss American Lutz argued powerfully that BMW was seen as a three-box sedan brand—especially in Germany, where it really counted—and that the home market would not buy into the idea of the 3-Series if it was a hatchback. Fresh in the minds of many executives must have been the humiliating failure of the '02 Touring, which had been very similar in concept to the 3-Series hatch proposal, and Lutz prevailed.

The spectacular sales explosion delivered by the eventual two-door, three-box 3-Series following its 1975 launch elegantly demonstrated the accuracy of Lutz's intuition, and it was two decades before BMW launched its next hatchback, in the shape of the 1995 Compact. As for Lutz, he went on to a glittering career in the auto industry, leading all three of the big US automakers.

BMW Classic
BFGoodrich
Radial T/A

E25 TURBO CONCEPT

The dramatic wedge shape that wowed the crowds, inspired the legendary M1, and nearly led to a partnership with Lamborghini

If evidence was ever needed of Paul Bracq's talent as a car designer, the E25 Turbo concept provides plenty. Very compact and low, with a strongly futuristic wedge profile, it came as something of a shock to conventional tastes, pre-dating Giorgetto Giugiaro's aggressive "folded paper" Lotus Esprit wedge by several years.

The Turbo was created primarily as a concept car to show off BMW's design and engineering prowess at the 1972 Olympic Games in the firm's home city of Munich, but the fact that it carried the E25 project designation shows there was serious production intent. The midmounted 2.0-liter turbo engine's 200 horsepower would have given it sector-leading performance at the time, and the advanced race-car style with gull wing doors would have made the planned two-hundred-unit production run a surefire sellout.

Yet, recounts Saxty in *BMW's Hidden Gems*, the CEO of Mercedes-Benz objected vehemently, insisting that the three-pointed star had exclusive rights to gull wing doors and that the Turbo should be scrapped. In any case, BMW had earlier sensed that the tide was turning against sports cars and toward protection for occupants and other road users. So, instead of focusing on speed and power, the Turbo was cleverly presented as a safety concept, its deformable front and rear sections highlighted in Day-Glo orange against the flame red of main structure. The Turbo was profoundly influential on BMW design, inspiring Giugiaro's spectacular M1 in 1978, the Z1 roadster in 1988, and the low-nose 8-Series the year after that.

Dramatic impact: Bracq's masterful 1972 Turbo concept inspired Giugiaro's M1 supercar as well as the Z1 and 8-Series. It used the 200 hp turbo motor but was billed as a safety research vehicle.

Too hot to handle: BMW built a one-off Compact with the mighty 321 hp engine from the M3 as a birthday present to a leading German car magazine. Would it have sold well? BMW clearly didn't think so.

M COMPACT AND M TOURING

Familiar shapes, but too powerful for their practical mission

There has always been a school of thought that the best recipe for a true performance car is to cram as much power as possible into as small and as light a body as possible. This was the theory tested out, not entirely seriously or with genuine production intent, by BMW M's renowned team of engineers when they slotted the massively potent 3.2-liter S50 Motorsport engine into the E36-generation Compact, a three-door hatchback little more than 4 meters in length and 1,200 kilograms in mass.

ABOVE: Rear view of one-off M Compact exercise shows the characteristic quad-exhaust setup, along with a small diffuser. Few outside BMW have driven the car.

Another M-car that never reached the showrooms. The Touring version of the acclaimed E46 M3 combined ballistic performance with palatial luxury and a roomy cargo bay. It took a further quarter century before BMW would market an M3 load lugger.

Ostensibly developed as a fiftieth birthday present for the influential German magazine *Auto Motor und Sport*, the M Compact was reported to have spectacular performance, hitting 100 km/h in a scant 5 seconds. However, it was destined to remain a one-off, though its rebellious spirit did find expression soon afterward in the M Roadster and M Coupe versions of the Z3 sports car and, much later on, in the luridly powerful 1M Coupe.

As for the M version of the E46 Touring, this was a much more serious proposition and, in hindsight, can be seen as a pioneering proposal for a compact "power wagon." BMW chose not to build it, but this genre soon took off as a big and highly profitable market, Mercedes-Benz and Audi calling the shots with their thundering V-8 and muscular turbo powerhouses. BMW has at last come around to this way of thinking, with the new and well-resolved M3 and M5 Touring models highly sought after. Even so, there must be a tinge of regret at not having got in on the act when it first came up with the idea.

Z1

The radical '80s roadster that could have been a racer, an off-roader, a dune buggy, or even a car-motorbike cross

BMW's 1988 Z1 (below) was more of a research project than a serious sports car. Exploring the potential of a steel chassis (bottom) topped with plastic body panels, designers came up with wild ideas including a multiterrain sports hatch (top) and even a two-seater dune buggy.

Developed in a record twelve months by BMW's then newly formed Technik GmbH creative skunk works, the compact Z1 roadster with its unique drop-down doors is a much-admired jewel in the BMW crown—even though it was only produced in small numbers. What is not so generally realized is that the Zukunft Technik subsidiary (for future technology) was actually a wide-ranging incubator and the Z1 was simply intended as a demonstrator for new technologies, in this case flexible plastic body panels bolted to a strong steel and aluminum chassis structure. The Z1 was not so much an individual model but more a way of thinking. That thinking was truly liberating for the many designers drawn into the Technik team, for it allowed a huge range of different body types to be built off the same chassis.

Many wild ideas ensued from the design team under Harm Lagaaij (who would go on to lead Porsche design for two decades), including a split-cockpit barchetta, a sports wagon that predated the M Coupe, and a spectrum of off-road designs that included a two-seater dune buggy, a rugged high-riding fastback (anticipating the later X6 and X4), and even carried the allure of a Paris-Dakar racer. But the wildest of all was Lagaaij's own idea of putting BMW's F1 championship-winning turbo engine, with its 1,000-plus horsepower, into a Z1 record-breaker. Despite some initial buy-in from R&D director Wolfgang Reitzle, this was quietly buried—and it would be many years before an F1-inspired motor would power a production BMW.

The Technik division itself would go on to mastermind a long series of concept and series production cars, including the Z1 and Z8 roadsters and the 1999 Z9 coupe concept.

BMW designers have been trying to recreate the magic of the 2002 ever since the late 1970s, and the Y2K project (inset, dropped in front of Madrid's Cibeles fountain in an early CAD image) targeted a year 2000 release but was superseded by the 1-Series. It wasn't until 2016 that Michael Scully came up with this 2002 Hommage tribute (main picture). Scully is now head of design at BMW M.

2002 REVISITED

The plan for a back-to-basics hotshot sedan to recreate that 2002 magic never got beyond the prototype stage, but its "think small" ethos led to the 1-Series

It's a safe bet that a good few hard-line BMW fans in the mid-1990s were looking ahead to the millennium and secretly hoping that BMW might pull something exciting out of the hat to celebrate the iconic 2002 tii in the year 2002—quite possibly with a fun small car that would recreate the buzz that had surrounded that breakthrough model in the '60s.

Well, at several BMW studios around the world—and within rival brands, too—there were eager designers doing just that, privately working up "take two" 2002 ideas as back-burner projects. Each was a major enthusiast for the original 2002 but had been working independently. To cut a long story short, their proposals eventually fused into a single fully drivable prototype, dubbed the 2K2, which was a compact two-door notchback built on an E46 3-Series platform.

The thinking had been very back-to-basics, with a stripped-back interior, lightweight construction, and a simple but powerful engine/transmission assembly. Performance was by all accounts impressive, and Steve Saxty recounts that top management gave its blessing to take the idea forward. However, there was a serious catch.

BMW was running into ever hotter water with its Rover acquisition at this time, and in a rare example of one project starving another one of funding, it was felt more prudent to invest the available budget into the potentially mass-selling Rover R30 five-door program than the much more specialized niche-market 2K2 idea. In the end, of course, Rover collapsed and was sold off, the millennium deadline sailed by unnoticed, and neither car got the green light. But the essence of the idea did live on: Some of its design themes showed through in the five-door 1-Series launched in 2004, and the rebellious spirit of the two-door notchback shines through in the chunky, truncated 1-Series Coupe that followed in 2007 (even though the designers had wanted it to be the first to launch, in order to establish a suitably sporting image for the whole 1-Series family).

WELCOME

A COMMUNITY BUILT AROUND A BADGE

BMW Welt, the futuristic brand experience center adjoining BMW's global headquarters in Munich, is a strong focus for BMW and Mini owners, car enthusiasts, historians, and school

Special exhibits and art, theatre and musical events are a regular feature at BMW Welt. Here a new iX3 is specially decorated to celebrate Valentine's Day in 2025.

The world might know the BMW Group as a globe-spanning organization with an array of premium brands, one that employs some 150,000 men and women in scores of factories, building 2.5 million cars a year and selling them through thousands of dealers worldwide. But it's so much more than that. The company and its iconic nameplates are at the center of a much bigger universe—the constellation of diverse concerns, associations, clubs, and individuals who drive these cars, who maintain them, fix and fine-tune them, buy and sell them, and, perhaps most vitally, love them so fanatically that their whole lives revolve around them.

In this chapter, we take a deep dive below the gloss and glamour of the showroom to speak to the people who deal with BMWs on a day-to-day basis: everyone from the service personnel who keep your car in good shape to the restorers who bring treasured oldies back up to concours standard, the auction houses that value and then resell those classics, the collectors who so avidly amass them, and, naturally, the massive fan base that lives and breathes BMWs, congregating in large numbers every year to swap stories, to size up everyone else's wheels, and to share their mutual passion for this most celebrated of marques.

THE FANS: THEIR FAVORITES, THEIR FOLLIES, THEIR FANTASIES

Wherever BMWs are sold in the world—and probably also in the very few territories where they are not yet marketed—there are fans who eat, sleep, and breathe the brand. BMW has become something of a religion to many of these enthusiasts, and the interest often goes far deeper than just the passing fascination of a chance purchase or the time-limited enjoyment of a BMW as a company car. For many, the intrigue soon proves so compelling that there can be no other possible choice of wheels, in the same way that a Manchester United fan could never contemplate supporting a different football team.

This worldwide undercurrent of brand fascination in almost every nation crystallizes in scores of clubs for BMW cars, Mini cars, BMW motorcycles and luxury scooters, BMW bike riders, and the owners of historic BMW Group vehicles such as the Isetta bubble car. Globally, there are eight hundred officially recognized BMW clubs, with almost a quarter million members—in Latin America alone, there are more than eighteen national BMW car and bike clubs. The BMW Clubs International Council acts as a link between these clubs and the BMW Group itself.

Most clubs operate on a local or regional basis, but at a national level they can be quite powerful and even exert influence on the parent company in Munich. As just one example, when BMW owners in the United States

Straight Six is the BMW Car Club UK's high-quality monthly magazine, and despite its catchy title it also covers models with three, four, eight, and twelve cylinders, as well as electric models and race cars.

Where else can you see such a captivating display of beautifully prepared BMW E9 coupes? BMW clubs all round the world meet regularly to show their cars, chat about all things automotive—and maybe indulge in some racing.

successfully put pressure on the BMW factory to add a six-speed manual transmission option to the E60-series M5 sedan, North American customers not welcoming the seven-speed sequential manual gearbox received by the rest of the world.

Running in parallel with the local and national clubs are subsections, or even separate clubs, for individual models. The BMW Classic and Type Clubs International counts thirty such clubs at the national level and more than twelve thousand individual members across five continents. These perform a valuable role in building a register of the vehicles owned by each member and providing owners with valuable advice and useful contacts for services such as insurance, maintenance, spare parts, dealing with commonly experienced problems, and technical support.

Another key function of these clubs is the social one. Regular newsletters and gatherings serve to bring members together to swap stories, experiences, and tips, and many organize group events, shows, and even track days and races for club members. This is where members can put in an almost fanatical number of man-hours preparing, fettling, polishing, and presenting their pride-and-joy possessions to other admiring devotees.

In this, BMW is no different from any other enthusiast or single-marque club, but members would naturally insist that their cars are a cut above the simpler, more humdrum models from volume-market brand owners' clubs. After all, what other brand has a near-110-year history that encompasses not only road cars and racing cars—including F1—but also motorbikes, racing bikes, microcars, aero engines, early mass-market family cars, and the latest and most powerful sports coupes on the market?

GETTING STARTED

"What first got you into BMWs?"

Putting this question to a selection of club members revealed a surprising range of answers. While many cited the formative early-childhood excitement of long journeys in

Pride of ownership: The 2002, seen here in US-market form with extended bumpers, is one of the models that often triggers owners into a lifelong fascination with the marque.

the back of a family BMW or admiring glances at the school gates, others gained their lifelong fascination later in their biographies. Technical fascination with BMW products was often a motive, while for others the entry point was enjoying BMW models as company cars—in some instances prompting the purchase of an older classic BMW as a project or a hobby.

One member even reported finding a tired-looking 2002 Touring in the early 1980s, long before the model became regarded as rare and desirable. Bought for a song, it provided weekend entertainment and an interesting contrast to his company car at the time, which just happened to be an E28 5-Series BMW. Another member—perhaps not entirely typical, as he is in charge of the BMW Car Club GB—outlined his automotive journey from fast Fords to VW Golf GTIs to BMWs. He has never wavered from BMWs since buying his first, an E36 318i sedan, and currently owns three, headed by a mighty V-12-engined M850i coupe—"a superb bit of kit," in his words. The roll call of BMWs he has owned and run over the years is impressive, encompassing an astonishing

The E28 5-Series is a model that many BMW enthusiasts aspire to, but the rare M5 version—which is vastly faster than its sober looks suggest—is at the top of many owners' wish lists.

Fast, fun, and flawed: The potent Z3M Coupe is a particular favorite among BMW enthusiasts: They're either aspiring to one or coming back for a second example. The open-top version seems less favored.

thirteen examples of the Munich brand: 318i, E36 323i Coupe, E36 328i Sport Coupe, Z3 M Coupe, E46 320i Cabriolet, E90 330d four-door, E92 335d Coupe, Z3 1.9i Roadster, F13 640d M Sport Coupe, F13 650i M Sport Coupe, F13 650i M Sport Coupe, E46 330Ci—and, of course, the big M850i.

Particular favorites, he says, were the Z3 M Coupe (he now hankers after another), the E36 and E92 Coupes, and, naturally, the current M850i. Any duds along the way? The answer is a clear no—perhaps the result of rigorously applying the advice he gives to others to "choose the very best car you can afford with a full service history and a car that stands out as being really well looked after."

GOING ONE BETTER

Impressive though that back catalog undoubtedly is, there is nevertheless one man who can upstage even this devotion to the Munich marque. Denmark-based Anders Bilidt is a self-confessed BMW nut—and the in-house expert on classic cars at leading auctioneer RM Sotheby's. "BMWs are a specialism of mine," he says. His professional analysis of the global market for collectable BMWs is given in the final section of this chapter, "The High-End Business."

In his private capacity, Anders is the most fervent of BMW enthusiasts, a fascination that has been with him virtually since birth. "My father co-owned a BMW dealership in Denmark from the 1970s," he says, "and I bought my very first car, a 1973 2002, at the age of sixteen. I still own it today—that's thirty-two years of uninterrupted ownership."

But the 2002 is far from the only car in his garage. It shares the space with a 1974 3.0 S, the big sedan that predated the 7-Series; a 7-Series some twenty years younger; and a 1997 323ti Compact—a model that some tip for future stardom. "I tend to prefer the smaller BMWs," he says. "Clearly the '02s, but I also love the E30 and E36. The E28 is quite special too, though."

And, confession time: How many BMWs has he owned and run since that 1973 2002? "More than I care to count," he admits. "Probably about ten '02s, but also the original Neue Klasse—E3, a couple of E21s, several E30s, two E36s, one E46, as well as an E12, E28, E34, and an E39—5-Series generations one to four, respectively—and an E32 7-Series."

What's missing from that list? Well, lots. If Anders's wish-list of BMWs is anything to go by, he will definitely need to call in the architects to triple the size of his garage. "I still

Entry ticket: Positive experiences with company cars, along with the model's solid reputation, make the E30 3-Series the favorite reason for investing in a BMW to enjoy. But hurry—prices are rising fast.

need to own a E24 6-Series coupe at some point," he pines. "But more importantly, I dream of adding a E30 325i Sport, an E36 M3, and a E28 M5 to my garage someday. A Z3 M Coupe would be great, too!"

Any mistakes or bad buys along the way? "None, really. Mostly, just failing to realize that I should have kept more of them."

And that sentiment, shared by almost everyone reporting back, is what BMW ownership is like. Especially when it comes to the smaller models of the generations covered in this book, the experience can be infectious, contagious, addictive—to the extent that ownership should perhaps come with a health warning that the taste of one model is likely to lead to an irresistible craving for more and more.

The cheapest and simplest models can provide a handy back door into low-cost BMW ownership. Entry-level models like this 316 Compact can be both fun and easy on the wallet.

THE PROFESSIONALS: TRADERS, FIXERS, RESTORERS, VALUERS, AND AUCTIONEERS

THE TRADER IN CLASSIC BMWs

While walk-ins to a glitzy BMW main dealer showroom are likely to be greeted by a warm handshake, a radiant white-toothed smile, and an avalanche of well-rehearsed high-pressure facts and smooth-talk sales spiel, the atmosphere at a nonfranchised BMW specialist is completely different. Perhaps the cars are crammed closer together, the spotlights are less intense, the carpets, if there are any, are less plush, and the coffee machines more antiquated; there could even be a faint whiff of oil in the air or the occasional mechanic in workshop overalls deep in conversation with a customer. Such is the honesty and informality to be found among the network of alternative businesses that deal in older BMWs and other cherished cars.

What is on show is not the latest factory-fresh design or a high-pressure "buy it now" deal on a slow-moving model. Instead, the fundamental motivation is a deep-down enthusiasm for the brand, underpinned by an even deeper knowledge of almost every model that BMW has produced since the 1960s. These specialized outfits don't just buy and sell collectible BMWs. Many of them employ dozens of skilled technicians who know the BMW back catalog inside and out and are responsible for looking after the hundreds of thousands of older models that owners choose to use as reliable and enjoyable everyday transport in preference to twenty-first-century machinery dominated by electronic driver aids and unfathomable touch-screen man/machine interfaces.

A prime example of this type of enthusiast-run business is Munich Legends in southeast England. One of the very first official BMW dealers in the United Kingdom in the 1960s, the place is now a thriving hub for all things classic BMW, from servicing, repairs, and rebuilds to sales, storage, and even events. The organization's Legends in the Fall gathering attracts scores of old and new BMWs and their owners to the Red Lion pub next door and is a regular fixture in the national BMW enthusiast's calendar.

Dan Norris, who has been running Munich Legends for the past several years, was originally a customer who was drawn to the business by the expert knowledge of the staff and enthusiasm of his fellow customers. Branding itself as "the ultimate BMW specialist," Munich's small showroom has at some time played host to almost every classic BMW in the repertoire—most especially the collectible M models and the big six-cylinder CSL-style coupes of the 1970s. But, says Dan, "we never sell more than ninety or one hundred cars a year. The engineering side of our business is much more important."

Dedicated independent BMW dealers staffed by knowledgeable specialists have a vital role in supporting the enthusiast community. This is Munch Legends in southeast England, one of the busiest in the business—and which gave valuable advice for this book.

One of four hundred: The M3 GT of 1994 was a homologation special with its 3.2-liter engine uprated to 294 hp. It's an aspirational car for many enthusiasts, and some are now rating it higher than the later E46 M3.

In terms of the cars they do sell from their small showroom, Dan is equally clear: "Our customers dictate to us what cars we handle. If they're not interested in it, then we aren't either." And right now, that interest seems to be E39 M5s, M3 CSLs, and, clearly, E30 M3s. "The problem with E30 M3s now is that they are all going to America," he explains. "Prices are going up, and Sport Evos are now all above £200,000."

Even regular E30 M3s—if any M3 can be described as regular—are pushing into six figures, whether the price is in euros, dollars, or pounds. There is evidence of some flattening off, but firmly on the rise are the 2002 tii and Turbo from an earlier era. These very rust-prone '60s and '70s survivors tend to be fully restored or need big money spent on them, with sound tiis topping out at perhaps £35,000 or £40,000, says Dan. Turbos are far rarer and fetch much higher prices: As of late 2024, Munich Legends was listing a smart example with under 50,000 miles on the clock for almost £100,000.

Overall, he says, the standard E30 is emerging as the car to have, and perhaps the best all-round value BMW classic to invest in: "It's a cultish car right now. Whether it's a saloon, a coupe, or a convertible, it's an easy car to own, and they're all good to drive." The Touring model is also one to watch, though its practicality is not as good as later generations.

In the broader picture, Dan observes, the E30's successor, the E36 generation, is beginning to recover after having dipped into clunker territory, with impecunious owners neglecting their vehicles and simply using them for transport rather than entertainment. Early examples are still very cheap but will require money spent; there is definite interest in the Tourings and, as always, the mighty six-cylinder M3, which is emerging from under the shadow of its universally lauded successor, the E46 M3—especially the ultra-rare CSL version. Even so, most agree that the "standard" E46 M3 is probably the best M3 to go for. It's the single model that appears most frequently at Munich Legends, whether for sale in the showroom or in for regular service attention in the workshops.

SOME DIFFERENT INSIGHTS

Barely an hour's drive from Munich Legends is another BMW specialist dealer, BMR Performance. Founded by former BMW main dealer workshop specialist Barry Sheward in 2018, it deals with all BMWs and Minis "from the 2002s to the big 7-Series." BMR doesn't buy or resell any cars but concentrates on servicing, restoration, and some bigger workshop tasks, such as engine swaps and auto-to-manual transmission conversions.

A self-confessed car obsessive, Sheward has owned "many, many" BMWs, including thirteen Alpinas and numerous M-models, and still provides a home to three E30s—which he says is his favorite generation of 3-Series. "What I like about E30s," he asserts, "is how well made they are, how well they drive, and how easy they are to work on for people who like a hands-on relationship with their classic car."

Barry speaks enthusiastically about a rare South African 333i 3-Series that BMR recently subjected to a full bare-shell

BMR Performance is another mecca for BMW enthusiasts, specializing in Alpina models, service work, and larger tasks such as auto to manual swaps. BMR also helped with this book.

Popular choice: As the second-generation 3-Series the E30 inherited the mechanical makeup of its predecessor—but also its 1970s-style fondness for rust. Good ones are getting rarer, but are worth seeking out. All versions are agile and fun on the road.

restoration. In its day, the South Africa–only 333i was in many ways a more affordable alternative to the M3, using a potent but easygoing 7 Series straight six for rapid but unstressed performance. Other officially modified E30s BMR has worked on include two Hartge models: the H27, with the big-block six, and the H36 M3 that somehow manages to shoehorn the M5's mighty powerhouse into the smaller shell. "They're very rare," Barry says, "and the many Hartge-specific parts are hard to find."

Of all the BMW-approved tuning companies, Alpina is by far the most widely known and respected—in fact, BMW Group took it over following the death of its founder, Burkard Bovensiepen, in 2023. Alpina was prominent among the semiworks racing teams that BMW supported in the glory days of the 1960s and 1970s, and today BMR Performance prides itself on being something of a regional hub for Alpina owners in the UK.

Beyond the tightly focused market for Alpina-style tuned BMWs, Barry notes that prices for the E36 generation now seem to have bottomed out and that a nice original six-cylinder manual example—a bit of a rarity, he admits—would make a sensible purchase. Already, he adds, the E46 is beginning to fall into the same pattern, prices dipping as the cars get older and less well looked after, but poised to rise again as soon as buyers willing to invest money to work on them begin to show interest. Overall, he explains, the E46 could be seen as "peak BMW"—more refined than the E36, with good style and image, a great cabin, and excellent engines. "They all drive well," he adds, "even the four-cylinders, and the diesels are very robust, too. I know of one which has gone for 300,000 miles without a rebuild."

As for the E90 range, from 2005 onward, they, too, are nice to drive, especially the sixes. But again, condition is important; according to BMR, some examples are already beginning to show signs of underside corrosion, the consequence of BMW using different coating solutions because of new chemical regulations that came in around that time. The engines are, in the main, robust, and even the top twin-turbo N54 3.0-liter six, acclaimed in the specialist press as one of the best performance engines ever, is strong and reliable despite its spectacular power and performance. The big proviso here is that the engine must be kept standard: In contrast to the older naturally aspirated power units, turbo motors are very easy to tune for even more power with a cheap and simple aftermarket remap of the control unit. This puts much more stress on all the components, leading to problems with the "bolt-on" ancillaries such as turbos and manifolds—or, worse, a blow-up that destroys all the internals. There are official BMW remaps, too, and competent dealers can immediately spot an unofficial one as soon as the vehicle is hooked up to the diagnostic machine.

EXPERT ADVICE

In terms of general warnings, both Barry Sheward at BMR and the technical people at Munich Legends are unanimous on many points when it comes to choosing and looking after a classic BMW. Firstly, and perhaps most critically, when BMW moved its engines over from cam belts to cam chains in order to eliminate the cam-belt replacement service action, the German company was being economical with the truth. Subsequent experience in service has shown that the cam chains do not last the life of the engine, as had been implied. The weak point here is the tensioner mechanism, which on the majority of models is made of plastic and wears away to leave the chain running slack or, worse, flailing around. Some earlier models, such as the E46, don't suffer as much, but wise mechanics will always check the engine oil for fragments of tensioner—a sure sign of impending trouble—and will replace the whole assembly every 100,000 miles.

Another more general warning relates to the surprising issue of corrosion. BMWs may be premium machines, but in the earlier years, the German company's rust-prevention measures were often no better than those of cheap volume-market brands. Most rust-prone of all are the '60s Neue Klasse sedans and the '02 series, but all models, right up to the 2005 E90 generation, are afflicted in some way. So, as general buying advice, look out for poorly executed accident damage repairs, and especially beware the car with a suspiciously thick and fresh-looking layer of underseal underneath—it's sure to be hiding structural corrosion and probably making it worse.

WORKSHOP WISDOM: BMW TECH SPECIALISTS TELL THEIR TALES

Speaking to a selection of seasoned BMW workshop specialists, a clear picture emerges of a group of highly motivated people. They have great enthusiasm for the products they attend to, but also certain very trenchant criticisms of the BMW parent company—most particularly when it comes to the below-the-surface quality of later model generations and the design philosophy that makes them harder and more costly to repair.

But just to be clear, BMW is not alone in any of this. Every auto brand, whether large or small, mass market or premium, has been forced to incorporate a whole host of incremental devices and control systems simply to stay abreast of tightening safety and environmental regulations and be allowed to market their cars. Rising costs at every level of production mean ever-more streamlined manufacturing processes that favor speed of assembly in the factory over accessibility for the workshop or home mechanic. Cars today are infinitely more complex than their 1960s and even 1970s forbears—just look under the hood of a 1960s

BMW 2002 and compare it with today's 3-Series equivalent, the 318i. The 2002 is laughably simple, its compact engine seemingly lost in the underhood void; the carburetor is clearly visible, as are the ignition system, spark plugs, distributor, air cleaner, and exposed exhaust manifold. Other than the brake servo, the radiator and its hoses, the wiper fluid bottle, and the hydraulic reservoirs, there's little else populating the space.

Contrast this with the modern 318i, where raising the hood reveals a mass of smartly finished black covers screening almost every single component from view. There are few clues as to what's underneath or even whether it's gasoline or diesel, though in some cases it's possible to see whether it is a six or a four. Removing the covers exposes a mass of neatly dovetailed control boxes, conduits, and trunking, but little in the way of the components familiar from kids' car books.

To the classical mechanic, this is an alien environment, the clear message being to stay away apart from doing token basic checks such as oil and water. Small wonder, then, that our independent BMW dealer technicians have to be real motor magicians—au fait with everything from simple spannering to obscure software issues resolved on a laptop screen. Fixing the enormous age range of BMWs means everything from antiquated plugs-and-points set-ups to today's powerhouses largely dominated by electronics; all this is way beyond the training given to modern-day main-dealer operatives.

Above all, experience is the key, knowing when to rely on traditional engineering and repair skills and when to reach for the plug for the diagnostic apparatus. What follows is a tiny fraction of the accumulated wisdom of our technicians, arranged model by model for an easier guide to the pleasures and the pitfalls of each 3-Series generation.

NEUE KLASSE SEDANS, 1961–1970

The car that saved BMW now needs saving itself. Though light and engaging to drive and mechanically robust, these classy '60s period pieces are structurally fragile and, like all cars of that era, very prone to corrosion. The few survivors are likely to have been heavily restored and change hands for substantial sums—but anyone lucky enough to find one languishing in a dry, well-ventilated barn could be in for an exciting and rewarding project.

As the car that saved BMW in the 1960s, the original Neue Klasse is highly sought after—none more so than the rare 1800 TI/SA, a race-crafted special edition that stormed Europe's circuits and laid the foundations for the M-cars that followed.

Precious asset: Very few cars from the '02 era survive, and those still in circulation have usually been expensively restored or rebuilt. Note the proud "automatic" script on the rear, from a time when manuals were the default option.

'02 SERIES, 1966–1977

This is the car that kicked off BMW's resurgence as a high-profile, fun-loving sports sedan brand—and, understandably, everyone wants one. Whether it's a humble 1600 or an energetic fuel-injection 2002 tii, it will be a lot of old-fashioned fun, but it will share the same shortcomings as the Neue Klasse from which it is directly derived. That means simple, solid drivelines and neat interiors, but also rust, rust, and more rust. Weak points are too many to mention, but a particular villain is the tailgate on the rare Touring: It dissolves completely and no spares are available, helping make the model even harder to find than it already was. Pricing of '02s is entirely dependent on condition and could range from zero to big money for a well-fettled 2002 tii or enormous money for an unmolested Turbo.

3-SERIES E21 GENERATION, 1975–1983

Though very different and much more modern in its style than its '02 series mentor, the E21 is almost 100 percent '02 technology underneath. So while it inherits the strong four-cylinder engines, decent transmission, and suspension of the older car, it carries over many of the same problems too. BMW evidently didn't learn much about corrosion protection between 1966 and 1975, for the E21 is equally rust-prone, a virus that has consumed many of the million-plus examples built. Keeping the water out is the key to survival, says Stuart at Munich Legends, with the rubber windscreen seals being a particular trouble area. The oldest of these cars have hit their half century, and unless an example is a "garage queen"—the dream purchase of a single-owner car that has been kept undercover for its whole life—it will have been restored. The small sixes are fabulous to drive, but many have been thrashed mercilessly, and early engines were prone to cracked cylinder heads. However, a four-cylinder 320 or 320i in good condition and regularly fettled by a specialist workshop could be a reliable daily driver that is both charismatic and entertaining. The asking price, as with all the oldies, will be entirely condition dependent.

E30 GENERATION, 1982–1994

The oldest E30s are past their fortieth birthday, so without question they fall into proper classic territory—and as proof of that, they now seem to be in vogue with serious collectors and prices are beginning to rise. The M3 is an entirely different market, with prices already in the stratosphere, fueled by clearly defined limited editions, restricted supply, and, of course, the model's historic status as the car that truly kick-started the whole M-car dynasty. As for the regular E30s,

It might be the entry-level model, but this first-gen E21 316 will still be light, agile, and fun to drive. But don't expect power steering, electric windows, or air conditioning—and if it's still roadworthy, it will be restored and probably expensive.

The car that came to define BMW: the 323i with its sizzling six and razor-sharp handling converted countless thousands to the BMW cause. It's the pick of the E21 bunch, but rust is its worst enemy.

most agree that they're more solidly built than their predecessors, but they still do suffer from the bugbear of corrosion. It's the biggest killer of old cars and can be found everywhere from the bulkheads to the sills, trunk, and roof, though the screens are bonded in and should not leak. The body is the most expensive thing to get right, observed one specialist, who warned that replacement panels are getting scarce, with many now discontinued.

That said, the mechanical elements are as strong as ever, with few provisos. There is a much wider variety of body styles than earlier models—the Cabriolet is a real classic—and there are even diesels and four-wheel-drives for the brave. Find a solid, well-maintained one and this epoch-defining model could turn a regular commute into a whole lot of fun; all of them drive well and handle tautly and entertainingly. The six-cylinder motors are creamy and characterful, while a particular delight is the agile sixteen-valve 318iS (late 1989 on) with an eager 136 horsepower propelling just 1,100 kilograms. Chosen carefully, any E30 could be a great introduction to the world of classic BMWs—and even a decent investment.

E36 GENERATION, 1990–2000

Here's where the BMW style changed, both inside and out. The E36's early quality niggles meant that it has continued to be undervalued—overshadowed by its clearly classic forebear, the E30, and its successor, the E46, held by many to be the best of the pre-electronics 3-Series. But the E36 has a lot going for it: Much of its technology is taken straight from the larger E34 5-Series, it is roomier than earlier cars, and there is the added glamour of a bespoke Coupe version and its visual affinity with the mighty second-generation M3, first of the six-cylinder superbreed. The Touring wagon is more practical, too, and the Cabriolet is just as classy and timeless as always. Corrosion is still something to look out for but is less of an issue than the previous generations, and the engines are just as strong and sparkling as ever; early sixes did suffer from the Nikasil coating issue, but most were sorted out under warranty.

All drive really well, with further improved ride comfort and handling—apart from one intriguing exception. The small Compact hatchback, introduced near the end of the run as a low-cost entry to the BMW brand, was heavily

Star of the future? As the third generation of the 3-Series, E36 models will soon be approaching classic status, which means that there may still be bargains among the clapped-out cars and clunkers in the classified ads.

critiqued on launch for its previous-generation dashboard and wayward dynamics. But it was steadily improved, and the much sportier final editions—318ti sixteen-valve and 323i six-cylinder—have the charisma and rarity to qualify as future classics. As for the rest of the lineup, they are now inching away from the banger stigma and into the classic waiting room, and as one specialist bravely ventured, it is hard to pick a bad one. Something to take seriously, then.

E46 GENERATION, 1998–2007

Just a few short years back, around 2020 perhaps, the E46 was *the* used 3-Series to buy. It combined the simplicity and clarity of the earlier generations with a newfound style and poise on the road, its cabin brought a luxury ambience and grown-up safety features, and the electronics had not yet become intrusive. The best of what are becoming known as the "analog" BMWs, the E46 generation models are now beginning to lose some of their shine as poorly maintained examples start to show bodywork stains and some mechanical problem areas at very high mileages. This is the generation that really brought diesel power to the fore, but some early direct-injection units now need expensive injector replacements; even so, well-cared-for examples will be faithful companions for many more years. The gasoline cars are universally good, and probably simpler and cheaper in the long run—and of course the Coupes and Cabrios are especially prized, and the Touring is both classy and roomy. BMW surprised the world when in 2001 it launched a second-generation Compact, this time based on the E46; this was a much more satisfactory machine than the E36 version, equal in specification and equipment to the full-size model, and the sportier engines ensured they passed muster as engaging, if low-profile, hot hatches. As with the E36 version, this could be one to watch.

On the minus side, a must-check point for any E46 buyer is the condition of the trunk floor. This is commonly thought to only affect M3s, but the experts say it afflicts the whole generation. The root cause is the rear suspension mounting directly into the rear box sections: The cracks in the floor panel that can result are expensive, and sometimes impossible, to fix. Not all cars suffer, however, and the E46 remains one of the best generations to go for—and the M3 is recognized as the best of all.

When the 3-Series grew up: The gen-4 E46 cars are more refined, with more of a big car feel. They make good family haulers, especially the Touring, but check the trunk floor for rust—it can mean the car's a write-off.

This potent M3 coupe version aside, the fifth-generation E90 models are the lowest in profile of the 3-Series back catalog. Longer lasting than earlier series, they're generally good to drive and could prove a canny long-term investment.

E90 GENERATION, 2005–2012

This was not a 3-Series that was universally lauded when it launched, and it never achieved the level of desirability that other generations did. Today, many examples are beginning to fall into the hands of owners unable or unwilling to spend enough on regular servicing, so they don't present the normally smart and shiny BMW image. That said, with E90 prices looking like they'll bottom out soon, something of a reevaluation seems to be taking place, and many are remembering the E90's qualities as a good driver's car, no matter what engine is up front. The Coupes and Cabrios have always been more sought after than the four-door sedans. It should be borne in mind that this was the era when the cam-chain tensioner issues began to be more widespread.

As for the M3 version—the only one with a V-8 engine—these are undervalued and now of increasing interest as other M3 generations climb in price. The big prepurchase check here is the state of the big-end bearings: It's a very high-revving engine, the tolerances are very tight, and failure can be terminal. The signs of trouble are excess noise on cold start and a noisy idle, and the cause is usually an owner who has regularly thrashed the car from cold. Yet, notes one expert, some run to 150,000 or 200,000 miles with no problem. Even so, he adds, it would be prudent to replace the shells if there is any doubt.

F30 GENERATION, 2012–2018

If the E46 was analog and the E90 semidigital, then the F30 is almost fully digital. Specialist dealers are now beginning to look after large numbers of these, and in general they are reported to be lasting well—as one would expect from a quality car less than fifteen years old. Where there are issues, say the specialists, they tend to relate to the engine ancillaries rather than the motor itself. For once, the diesel's

What will history make of brand oddities like the 3 Grand Turismo? Non-mainstream models like the Z1, the 2002 Touring, and the two Compact generations have come to be viewed with great interest—and not just because they are rare.

bill of health seems to be slightly cleaner, with gasoline cars sometime suffering injector problems and, of course, the timing chain tensioner—betrayed by a noisy start-up from cold. Almost every model has one or more turbos, so there may be hidden dangers in examples that have been unofficially chipped or remapped. It is early days for this generation, of course, but one thing is already certain: The preponderance of electronic systems in almost every facet of the vehicle means that well-meaning enthusiasts and home mechanics are unlikely ever to get a look-in.

HIGH-END BUSINESS: AN RM SOTHEBY'S EXPERT ON AUCTIONS, VALUES, AND EXPECTATIONS

In 2022, at a closed auction in Germany, a 1950s Mercedes-Benz went under the hammer. The tone was immediately set when the opening bid came in at €50 million—already way in excess of the top price ever paid for that most vaunted of Ferraris, the 250 GTO. And when, amid mounting tension in the room, the hammer finally fell, the winning bid was a staggering €135 million—by far the highest price ever paid for a car anywhere in the world. Auctioneer RM Sotheby's observes that this car now sits among unique works of art in the top ten most valuable items ever sold at auction.

It helped, of course, that the bidding took place in the Mercedes-Benz Museum in Stuttgart, and that the car, in perfect condition as part of the museum's collection, was one of only two examples ever built. The 300 SLR "Uhlenhaut Coupe" prototype, named after its creator and the chief engineer behind the triumphant straight-eight Grand Prix Mercedes racing cars, is in effect a Formula One car in a dinner jacket: Strip off the tuxedo and it is pure F1 underneath. Its nearest equivalent today would be the Red Bull hypercar or the AMG One, also from Mercedes. The ultra supercars built by McLaren, remarkable though they are, don't even come close.

Even the quickest glance at the ten most valuable cars sold at auction gives the impression that the top-price contest is a stitch-up. Apart from the record-breaking SLR and another Mercedes in seventh place, all the other cars are Ferraris. The highest-placed other marques are a handful of

In contrast to other premium marques such as Ferrari and Mercedes-Benz, BMW has few of the universally recognized brand heroes that help ratchet up auction prices. Among older BMWs the closest to celebrity status are the 328 roadster (above) of 1938, and the mid-'50s 507 (top) has enough glamour factor to break into the seven-digit auction price bracket.

Aston Martins, Jaguars, and McLarens in the lower reaches of the top twenty. But of BMW there is no sign whatsoever—until 275th place, where a 507 sold in 2018 for $5 million finally appears.

So why is this? BMW has a much longer history than any of these other carmakers, yet collectors appear to have eyes for only Ferrari, Aston Martin, and the like. Why is BMW missing out?

People collect because that's where their passion is, says the European car specialist for RM Sotheby's, Anders Bilidt: "Ferrari, for instance, has a huge sporting heritage, and a brand like BMW is never going to be able to compete with a marque which is purely identified with sports and racing cars—they fuel the passion."

But while the spectacular Uhlenhaut Coupe result may be symptomatic of a new trend toward cars as pure art investments, in the same way a van Gogh or Cézanne might be, most buyers of big-ticket classic cars are motivated more by fascination with the cars themselves than any sense of investment for financial gain. Many of them are building private car collections of their own—and some even have their own test tracks so they can drive and enjoy their possessions away from the public gaze. Hundreds of wealthy collectors fan the flames of the transaction prices as they seek to outbid one another for rare Porsches, Astons, Lamborghinis, and, of course Ferraris. But once again, there appears to be little high-profile interest in BMWs.

WHAT HAPPENED TO THE BRAND HEROES?

"BMW as a whole hasn't quite achieved the same recognition that, say, Mercedes-Benz has," explains Anders. "It's the case everywhere—sports cars, racing cars, they have the history, they get all the glory. Sports cars are always preferred as collectors' cars—it makes sense." And it is the reflected glory of these halo cars that helps fuel the interest in the humbler models lower down the value scale.

Come the 1970s and BMW was on a roll, toying with the idea of challenging Porsche and Ferrari on the racetracks and even having a crack at Formula One. One by-product was the fabulous M1 supercar, which at last put BMW in the sights of serious car collectors.

Apple cofounder Steve Jobs wasn't a car enthusiast but loved the simplicity of the BMW Z8. His example sold for $329,500 at a RM Sotheby's New York Icons sale in 2017. In the same sale eighteen cars sold for more than $1 million, a Ferrari for $17.9 million. The VW Microbus made $207,000.

High-profile BMWs with star status are few and far between; the prewar 328 sports car, for example, might have been an engineering groundbreaker, but it's not a car that most people would instantly recognize. By contrast, both the 1950s BMW 507 roadster and the 1980s M1 do have that inbuilt glamour factor, the visual charisma that elevates them beyond the reach of midlevel collectors who mainly shop for technical fascination or historical significance—Elvis drove a 507. And both 507 and M1 have the essential ingredients for top-flight collectability: They look fabulous and they are extremely rare, as they were both commercial failures in period, leading to production runs under three hundred units. That lifts them into the investment category—but what of regular BMWs that mere mortals can afford? Models for which the main enjoyment comes out on the road rather than in watching the value rise?

The nearest thing to a halo model in BMW's modern catalog is the E30 M3 from the late 1980s. Top auction prices paid for the most desirable Sport Evo version hover in the high $200,000s, but, reinforcing its still-accessible status, there are plenty of M3 bargains available for one-tenth of that figure.

THE DRIVER'S CAR—DRIVING VALUES

BMW may not be able to boast the same repertoire of value-fueling brand icons as Mercedes, says Anders, but the fact that its models are not yet seen as investors' cars could be a blessing in disguise. There are between ten and twenty prominent BMW collectors worldwide, compared with hundreds for the exotic Italian marques, so prices are still within relatively easy reach.

That said, he points out, the signs of price escalation are beginning to appear, and the 1980s M3 is moving well into six-figure territory. "This '80s icon competed with the Mercedes 190 2.3 on the racetracks," he explains. "It won more races than the Mercedes, and now it's achieving higher prices too."

Historic BMW icons such as the 1920s Dixi are highly prized but do not yet command high prices, and the same holds true for an ultra-modern BMW—the i8. "It's a design that stood out as it was very different, and it was built in low numbers," says Bilidt, "but only recently have we seen values begin to rise."

COUNTDOWN TO CLASSIC CELEBRITY

As a lifelong BMW fan, Anders Bilidt of leading auctioneers RM Sothebys watches BMW values closely, but most don't appear on his professional radar, as few models yet meet the €100,000 threshold above which the majority of RM Sotheby's clients do their shopping. But privately, he takes a passionate and almost forensic interest in what's bubbling under in the lower value bands. Here are his comments on some of the key models in the history—and prehistory—of the 3-Series ranges covered in this book.

- **1960s Neue Klasse and '02 series:** "There were a good few built, but they all rusted, and many have been crushed; they could be an investment, though values are going down as well as up. They are collectors' cars in their own right—the 1800 TI/SA fetches big prices, the '02s are getting stronger, especially the Turbo, and early Baur full convertibles are sought after."

- **E21:** "These were cheap for a long while, but the 323i is definitely a collectors' car and not difficult to find. Those still left tend to be in good to great condition."

- **E30:** "Anything M3 is a collectors' car, and eventually the more mundane models will be of greater interest, too, though they will never see the same percentage increases. BMWs are actually driven more than other classic cars, whether it's a 320i or an M3, so buy the best you can afford. An E30 Touring could prove a better investment than a four-door—it was quite special when it came to market."

- **E36:** "These are edging into classic territory now, though not often used as daily transport. The M3, 325i, and 328i have become 'Sunday cars,' and the convertibles have their own little niche as cruisers, never driven hard; they can attract a premium. A 323ti Compact could be a good one if unmolested—it's something out of the ordinary."

- **E36 M3:** "I actually prefer this to the E46 version. It's the most undervalued M3, the last generation before electronics set in, and it has a stunning engine. A hidden gem is the E36 M3 GT, in British racing green—values are coming up, and it's the one to have."

- **E46:** "My personal opinion is that this is not the best 3-Series. The interior got cheaper, the electrics are not as good, and the quality was better on the E36."

- **E90:** "Only the M3 with the V-8 engine stands out here, though it didn't hit the mark at the time, as it was too quiet with the standard exhaust. Overall, it's not as involving, as there are more driver aids—though you can still switch them off, and an original, unmolested example could be quite special. Collectors are definitely looking for analog cars, and this one is definitely more analog than later models."

- **F30 and G20:** "I can understand why these aren't yet so favored: The M3 and M4 are too heavy, they are turbocharged, and there are too many driver aids, so they're not as responsive to the throttle. Die-hard fans don't know what to make of it. Limited-edition versions are not necessarily investments but could at least hold their value. Here, we have to reset our criteria about what is special. Now it is production numbers, so investors will be looking for those limited editions with low production numbers, low mileages, and unusual colors—not black or silver."

Overall, Anders concludes, people who choose BMWs tend to have different values from other buyers: The driving experience is the top priority. And here, he says, the E30 M3 is truly the ultimate driving machine, but the E36, the E46, and even the E90 can also be in the running.

"But after that," he says with a sense of foreboding, "different values seem to apply. Nowadays, cars are anything but analog. They are a comfortable couch to sit on—and for Sunday cars you want the exact opposite.

"BMW enthusiasts like me want manual transmissions, naturally aspirated engines, and rear-wheel drive—the old-school driving experience. Whether they have €20,000 or €2 million to spend, BMW fans want the same thing. That's what it's all about."

And who could possibly disagree with that?

EMBRACING ELECTRIC: THE BOLDEST ADVANCE OF ALL

As the 3-Series moves into its eighth, now-electrified generation, BMW has given a glimpse of the radical new design language that will eventually characterize all future models. Gone are the assertive forms and elaborate surfaces, and instead comes a much simpler, almost purist

BMW

OPPOSITE: Pure and simple: BMW's fresh design aesthetic for the electrified era sees clear, crisp edges and clear planar surfaces with gentle intersections. Rather than a riveted-on badge, the BMW emblem on the hood is etched into the surface of the metal.

ABOVE: BMW design sketch for the Vision Neue Klasse, which previews some of the themes of the upcoming eighth-generation 3-Series, shows clearly the use of light as a decorative element, with the grille area even able to project messages.

At the time of writing this, BMW has yet to reveal the full detail of its much-hyped Neue Klasse project, the all-embracing electric vehicle architecture that will eventually underpin a huge array of BMWs of all shapes and sizes.

The unveilings will begin with the replacement for the iX3 in 2025, with the most momentous announcement—that of the brand-new successor to the 3-Series—held back to 2026. And for good reason, too: 2026 marks fifty years since the first generation 3 appeared in the showrooms, an anniversary of great significance to the company. Mindful of the central importance of the 3-Series to almost every aspect of its business, BMW will want to make as big a splash as possible with its breakthrough new model dynasty. So, quite understandably, key technical details will be kept under wraps until much closer to launch date, which is something of an issue for publications coming out in advance of that landmark to celebrate it.

Even so, there is a lot we do already know. The Neue Klasse–based 3-Series, the next chapter in BMW's super-successful model dynasty and the eighth since 1975, is likely to be the most pivotal of all the eight generations for one very simple reason: It will be the first ever pure-electric 3-Series and, in true BMW style, it will launch a whole raft of new technologies. Expectations are sky-high, but as an engineering innovator rather than a follower, BMW is unlikely to reveal much in advance.

Nevertheless, as we explore in this chapter, thanks to some carefully placed comments by senior company figures and insiders speaking off the record—and, of course, the four Vision concept models from 2021 to 2024 providing strong visual cues—we can piece together much of the picture.

But first, a short primer on BMW's electric history.

LEFT: The premiere of the Vision Neue Klasse concept in 2023, including (fourth from right) BMW CEO Oliver Zipse, who has always believed in developing electric and combustion-fueled models in parallel.

BELOW: Front view of the Vision Neue Klasse concept car, showing the use of lighting elements to create a three-dimensional effect in the grille. SUV models in the new series have a very different interpretation of the double kidney feature.

Such was BMW's confidence in spreading the message of its new ideas that it took the Vision concept to the 2024 Goodwood Festival of Speed. Sizable crowds gathered around the enclosure to weigh up the fresh new look.

AMPS, VOLTS, AND OHMS: HOW WE GOT HERE

Even though it is proud to have Motor as its middle name, BMW has always been open-minded when it comes to technical solutions that might promise better performance, increased efficiency, lower emissions, or reduced demand for scarce natural resources. Way back in 1972, a battery-powered 1602 led the marathon race at the Munich Olympic Games. In the 1970s and 1980s, BMW pioneered high-performance diesels and began looking at hydrogen as a motive fuel. In the 1990s, the popularization of highly efficient diesels helped slash fleet CO_2 emissions. And the early 2000s saw the special H2R hydrogen-fueled streamliner break a series of international speed records.

The group underlined the seriousness of its R&D in battery power in 2008 by building a batch of five hundred special electric Minis for a massive field trial in North America and Europe. This was followed up with further testing exercises, this time in China, too, with the ActiveE, an electric version of the 1-Series Coupe.

Around the same time, BMW finally confirmed rumors that had been circulating for a couple of years that the company had set up a special skunk works of top-level engineers and blue-sky thinkers to take a fresh look at the broader issues of transport and the environment, especially in the growing numbers of sprawling megacities, most notably in southeast Asia. The initiative took on the name Project i, and its free rein when it came to vehicle design resulted in the highly advanced i3 city car in 2013 along with the decision to manufacture the i8, the world's first plug-in hybrid supercar, derived from the Vision Efficient Dynamics concept that been such a sensation on its unveiling in 2009.

Following the i3 and the i8, BMW's electric journey began to accelerate with plug-in hybrid versions of key models, such as the X5 and the 3-Series' 330e, plus the first full-electric iX3. The journey is best summed up in these four phases:

- **Phase zero, 2008–2010:** Data-gathering field trials with Mini and 1-Series adapted for battery power.
- **Phase one, 2013:** Advanced i3 city car with radical design, aimed at early adopters; i8 supercar to establish performance-car credibility; plug-in hybrid versions of regular models, such as 3-Series and X5.
- **Phase two, 2019:** Battery-powered versions of mainstream models, such as 4- and 5-Series and X3, that look like the regular models; small to medium volumes, not staple products.
- **Phase three, 2025–2026:** Big-volume core models where electric power is at the heart of the range as the mainstream choice, not a minority option; designed around exclusively electric architectures; new iX3 and 3-Series will be the first of these.

VOICES OF DISSENT

The repercussions of Russia's invasion of Ukraine, along with steep rises in the cost of living and crowd-pleasing populist policies in some nations, have resulted in the relaxing of environmental targets in many European countries. This comes despite increasingly apocalyptic warnings from the United Nations and other climate groups. The main backlash has been against electric cars (EVs), just as the volume manufacturers were gearing up their supply chains and manufacturing networks to transition from combustion to battery power. The upshot: The EV no longer looks like such a done deal for the future.

But BMW appears to be staying true to the multifuel "power of choice" agenda that it has been pushing for a decade or more, even though the firm sparked the fury of climate-change hardliners and some investors for its refusal to name a final phase-out date for its combustion-engined vehicles. It argues that there will always be markets with minimal charging infrastructure where combustion cars or plug-in hybrids are the only option.

The sporty i8 and the urban-focused i3, both shown here in their early auto show concept forms, were a strong pointer that BMW would pursue electric power through innovation and originality rather than by adapting existing products.

ABOVE: BMW resorted to costly carbon-fiber construction to ensure its 2013 battery-powered i3 city commuter was light enough to give the agile driving characteristics expected of a BMW. It was a true trailblazer in both style and engineering originality.

RIGHT: The year is 2008, and BMW surprises the industry with an ambitious field trial of 600 Minis fitted with electric drive trains and a traction battery in place of the rear seats. The study gave BMW valuable feedback on the technology and how people actually used their vehicles.

ABOVE AND OPPOSITE: BMW followed up the 2008 electric Mini field trial with a larger exercise using electrified 1-Series coupes. Thanks to improved batteries and the model's larger dimensions, the ActiveE enjoyed better range - and a rear seat for two passengers.

Right now, that multienergy strategy might seem to have been a sensible choice. But with BMW bosses stating on several occasions that the next generation of medium-sized cars will be exclusively electric, where does that leave the new 3-Series?

After all, the 3-Series is BMW's heartland product, the car that the company is most closely identified with; it is the upholder of the brand's global identity as well as its biggest seller and a principal driver of its turnover and profits. Industrially, commercially, and reputationally, it is clearly a very big bet to replace a signature product with something so completely different. It's as massive a reset as the one Volkswagen made in the early 1970s when it replaced the Beetle with the Golf. But BMW's action, though it is sure to seem smoother and less disruptive on the surface, could prove the more significant in the long run.

TWIN-TRACK APPROACH

It is already clear from the three showtime prototypes displayed in 2023 and 2024 that the company is continuing full speed ahead with its Neue Klasse EV technology toolkit and that the first vehicle built on this new architecture is appearing in 2025, with a further six variations set to debut in successive years. There's even a brand-new factory in Debrecen, Hungary, dedicated to these new-generation models.

The Vision studies give a very clear idea of what the series-build new-generation models could look like, and insiders who have seen the finished production versions confirm that they are very close to those design studies. A word here about the naming protocol for BMW's many design studies: Hommage models are modern

reinterpretations inspired by classic models from BMW's history. Concept models act as "softeners" for proper production cars set for imminent sale. Vision vehicles point toward the near future and give a strong steer about the direction of design thinking, rather than a blueprint for an actual model.

So in the case of the i Vision Dee from January 2023, the Vision Neue Klasse from September of the same year, and the Vision Neue Klasse X six months later, we can get a strong sense of the overall proportions and form language of the upcoming production models if not hard detail on the final design or engineering. The new-generation models look much fresher and purer in their design language, with spacious and airy passenger compartments, big windows, and futuristic driver interfaces; the X models will be more rugged looking, but without the complexity or polarizing attitude of current designs.

Yet despite this massive investment in the electric architectures, it has become clear in recent months that BMW is also planning a parallel track of combustion-fueled alternatives to help it ride out the temporary anti-EV backlash. Understandably, though, several unknowns still remain about the full picture of the new generation—and these boil down to five principal questions.

1. What will be the sequence of model releases?
2. Will they be battery, combustion, or both?
3. What will the production versions look like, and will the combustion cars emulate the new EV design?
4. What will power the next generation of M3 and M4 high-performance derivatives?
5. What will happen in the longer term to the combustion-engined cars?

WHAT WE KNOW

1. WHICH MODELS WILL WE SEE FIRST?

The replacement for the current iX3 will be the first model to employ the Neue Klasse electric architecture and is set to go public in 2025. Next up will be the eighth-generation 3-Series, which may be previewed or even unveiled at the Munich auto show late in 2025 and is confirmed for commercial sale in 2026. A Touring station wagon is expected to follow closely, as is the first of the new 4-Series, most likely the Coupe. The new M3 and M4 (see #4) are penciled in for 2027–2028.

2. WILL THE NEW 3-SERIES BE POWERED BY BATTERIES, COMBUSTION ENGINES, OR BOTH?

The short answer now appears to be both. The new Neue Klasse–based 3-Series will be electric in all its forms, but it will be paralleled in most instances by gasoline- or even diesel-powered versions developed from the combustion-centric CLAR architecture that underpins all of BMW's current medium and large models. Just as with today's cars, both battery and internal combustion ranges are expected to offer the choice of either rear- or all-wheel drive.

3. WHAT WILL THE NEW 3-SERIES LOOK LIKE, AND WILL THE EV VERSIONS BE VISUALLY DIFFERENT?

It is already clear that the new models will mark a major step change in style from the current generation. They will be smoother, simpler, and purer than today's cars, and their proportions will be different—the hood can be shorter, as it no longer needs to accommodate the length of a straight-six combustion engine. The big question is this: Will BMW update the look of the combustion-engined cars running in parallel with the EVs to match the style of the fresh generation? The likelihood is yes, as BMW already has a strong

An early design sketch for the Vision Neue Klasse X concept reveals a more aggressive form language than the calm, relaxed approach of the sedan version. This X concept serves as a blueprint for the new iX3 launching in 2025.

track record of doing just this. The current iX3, for instance, is a very different car from the combustion version, yet they look identical in size, proportion, and style even though detailed differences signal which is which.

4. WILL THE NEXT-GENERATION M3 AND M4 BE ELECTRIC TOO?

This is an issue of the utmost seriousness for hardline M-fans, who revel in the fabulous performance delivered by the twin-turbo S58 straight six. With the latest edition of the larger M5 having just moved to plug-in hybrid power, the fear in enthusiast communities is that the M3 and M4 will go all the way and become fully electric, thus closing the door on the era of high-powered gas-fueled enjoyment.

Those fans have no need to worry—yet. M Division boss Frank van Meel has confirmed that although the priority in the development of the new M3/M4 on the Neue Klasse architecture is the battery-electric version, there will continue to be a gasoline-powered range, too. This remains based on the current CLAR platform but with the power train reengineered to meet the stricter emissions regulations coming into force before the end of the decade. For further details on these models, see the sidebar on page 210.

5. WHAT WILL HAPPEN TO THE COMBUSTION-ENGINED CARS IN THE LONGER TERM?

This is a tricky one to call. The simplistic answer is that it is mainly in the hands of the legislators in the major economic blocs—but also in the hands of the chemists and inventors in the leading fuel companies and universities. Several European nations have legislated that the sale of new gasoline- and diesel-engined private cars will be outlawed by 2035 or 2040, but there is still intense debate on the role of plug-in hybrids and to what extent they qualify as electric. Further debate (especially in Germany, where high-powered cars are seen as central to the country's economy) surrounds the issue of synthetic fuels. Lobbyists argue that if synthetic fuels can be made carbon-neutral when used in combustion vehicles, then new models running on these fuels should continue to be sold.

BMW's argument is that it will continue to build combustion-engined cars for as long as there is demand for them. In contrast to Volvo and several other major automakers, the company has declined to name a date for the phasing out of gasoline and diesel car production—so the likelihood is that these models will continue until there is a worldwide ban.

NEUE KLASSE TO NEW 3-SERIES: HOW BMW'S THINKING EVOLVED

In the first decade of this century, it's fair to say that most automakers were in denial about the need to shift away from fossil fuels and toward energy sources that emitted fewer—or no—climate-forcing greenhouse gases. Few dared contemplate the industry-wide upheaval that would accompany such a transition; instead, the focus remained on abating the tailpipe pollution from existing gasoline and diesel engines. Highly successful diesel technologies and exhaust after-treatment systems led many to think that the problem had been solved, or at least pushed many miles down the road.

But BMW, as we have already seen, was one of the few majors that chose to dissent and take the issue of fossil-free power seriously. Its Project i program provided the technical expertise and consumer recognition with the pioneering i3 and i8 to take battery power forward into a broader customer base. Engineering advances enabled partial electric power—in the shape of mild and full plug-in hybrids—to be added to existing mainstream models, but only relatively recently have fully electric versions of current cars been rolled out.

Yet as every engineer working on those models readily admits, adapting a combustion-oriented vehicle architecture to full electric power is riddled with compromises. Not least of these is the need to provide a large flat space for the batteries within the chassis envelope, set as centrally and as low to the ground as possible in the interests of weight distribution and handling. The engineering consensus has always been that a dedicated EV platform is required to maximize the benefits of battery power; the commercial counterargument once held that it would be far too costly to develop two separate platforms to take BMW into the future.

Whatever BMW may have already decided internally over those years, the debate among external commentators and industry watchers still revolved around which way BMW might eventually jump. Would it continue to evolve its Cluster Architecture for both combustion and battery power, would it switch over to a fresh dedicated architecture purely for electrics, or would it do both?

(CONTINUED ON PAGE 212)

Designers and modelers at work on the fine detail of the Vision Neue Klasse X concept clay model in the studio.

POWERING THE PARADIGM SHIFT

BMW BELIEVES BUYERS SHOULD HAVE THE POWER OF CHOICE, SO THE EIGHTH-GENERATION 3-SERIES WILL OFFER BATTERY AS WELL AS FOSSIL-FUEL DRIVELINES.

The dictionary defines a paradigm shift as "a fundamental change in approach or underlying assumptions"—and this is precisely what BMW is intent on achieving with this fresh-generation 3-Series.

That big change in approach has already been successfully accomplished on an engineering level, with battery-powered versions of several BMW model lines outperforming their gasoline- and diesel-fueled parent models (witness the i4 M50's spectacular 0–60 mph acceleration of 3.7 seconds). Yet, for many—especially often-cautious premium car buyers—underlying assumptions and attitudes regarding electric cars are much harder to move. Many have heard of Tesla and other e-power pioneers but continue to believe that electric cars are dull and have limited ranges before needing to be recharged, that the charging process takes several hours rather than mere minutes, and, with a grain of truth, that in many countries there are few—or even zero—public charging points. The view could be summarized as this: Electric cars are not for regular mass-market drivers but instead purely for virtue-signaling devotees.

BMW's mission with the new Neue Klasse–based 3-Series is to change those outmoded views and to prove that in the rigors of everyday use, battery power in a midsize car is just as convenient and effective as fossil-fuel energy. To reinforce the point, BMW will market the new car with the option of a wide spectrum of drivelines that include pure battery power, putting the choice in the hands of its customers.

Those models will look almost identical—and a lot like the Vision Neue Klasse concept pictured. But underneath they will be chalk-and-cheese different, with the electric models being built on the brand-new, purpose-designed Neue Klasse architecture, and the fossil-fueled models employing an updated version of the Cluster Architecture (CLAR) structure of the current G20-generation models.

So how, exactly, do these look-alike cars differ under the skin?

The twentieth-century science-textbook picture of a fossil-fuel powertrain is a simple one: gasoline tank, carburetor, pistons, clutch, gearbox, driveshaft, axles, and wheels. But that's barely recognizable in a modern car, which has grown infinitely more complex over the years. In the name of cutting toxic emissions and making the vehicle easier and safer to handle, cars have sprouted so many add-on devices and systems that the core engine is dwarfed by all the surrounding equipment needed to tame its wilder tendencies.

By contrast, an electric drivetrain is astonishingly simple. At its most basic, it's a battery that feeds current to an electronic inverter that transforms the voltage to the right value for the motor. From the motor the drive goes straight to the road wheels, occasionally via a two-speed gearbox on faster models. No clutch is required as the e-motor produces its full torque as soon as it starts turning. It can become more complex in sportier models with multiple motors and all-wheel drive, or when the battery has to be capable of accepting very high charging voltages and currents, or when it is also tasked with feeding energy back to the power grid in a bidirectional charging arrangement. Yet at no stage is it anywhere near as complicated as a fossil-fuel setup of equivalent performance.

Industrially, it is proving to be a significant step, graduating from the established processes of engine manufacturing to an electrical environment that has more to do with clean precision engineering and electronics than the heavy industries of the nineteenth and twentieth centuries. Vehicle layouts are having to adapt, too, principally to provide a central position low in the chassis for the battery, and to engineer the vehicle to cope with the extra weight that current batteries bring.

LEFT: The original Neue Klasse sedan in the light tunnel where paintwork and finish are meticulously examined. By today's standards it looks tall and narrow, but still very clearly a BMW.

BELOW: BMW gen6 electric drive units on the assembly line at BMW Group Plant Steyr in Austria. The clean-room conditions contrast with the noisy environment of combustion engine manufacture.

ABOVE: The BMW gen6 battery uses the underfloor space and the move from prismatic cells to 800-volt cylindrical technology helps give a 20 percent boost in range, to 800km.

LEFT: The low and wide frontal silhouette of the Vision Neue Klasse concept in the same light tunnel. The twin BMW grilles are now stretched fully outwards and can incorporate a variety of patterns and lighting signatures.

OPPOSITE: BMW CEO Oliver Zipse presents the Vision Neue Klasse concept car at the 2025 Shanghai auto show.

ULTIMATE *ELECTRIC* DRIVING MACHINE

PETROL HEADS, WORRY NO MORE: THE NEXT-GENERATION M3 WILL COME IN TWO FORMS: ONE WITH THE CELEBRATED SIX-CYLINDER PETROL ENGINE, THE OTHER POWERED BY BATTERIES THAT PROMISE TO SUPERCHARGE THE DRIVING EXCITEMENT.

Height of ambition: BMW's VDE prototype scales a near-vertical wall at the Shanghai auto show to demonstrate its unprecedented traction and grip.

BMW listens to its customers, none more so than the fanatical, often strongly opinionated owners of the firm's high-performance M cars. Recognizing the strong concern among this enthusiast clique that the raw thrills and intoxicating soundtrack of highly tuned combustion engines might be forever lost in the transition to electric vehicles, the company is softening the blow by running both forms of power in parallel versions for the next generation.

BMW's confidence in its new technology is such that it believes customers will see for themselves the remarkable things the new architecture can achieve and will make up their own minds about which engineering to opt for.

As if to prove the point with maximum drama, the company recently showcased the Vision Driving Experience (VDE) technology demonstrator, a Neue Klasse–based vehicle packed with all the company's very latest engineering thinking, much of it still secret and kept away from reporters' prying eyes.

One headline-grabber is the powerful built-in fan whose job is to suck the car toward the ground to increase cornering grip, stability, and braking power to unprecedented levels. This is unlikely to feature on the production version; indeed, BMW stresses that the VDE demonstrator is not the new M3.

What BMW does say is that the VDE features four electric motors and the ten-times-faster electronic superbrain control units that will enable exceptionally rapid responses to inputs and signals. In particular, this allows the braking system to operate almost entirely through energy regen, with much less reliance on conventional (and energy-wasting) friction brakes. One result is a claimed 25% boost in efficiency, which on road-going versions would translate directly into increased range on each charge of the battery.

As for the battery itself, the whole Neue Klasse line will use advanced 800-volt systems for faster charging and lighter cabling, and BMW's M division is known to be working on enhanced versions of the batteries to unlock even more power at the wheels. Figures of up to 1,000 horsepower have been cited in connection with possible M versions, and the explosive performance of the VDE certainly bears that out. At the Shanghai auto show the prototype was able to climb a seemingly gravity-defying near-vertical wall.

Overall, it is clear that BMW has all the elements in place to extend the performance envelope still further when it comes to the outer limits of handling, grip, stability, and braking, as well as outright linear acceleration. And by choosing an electric edition, tomorrow's M-car customers will be the first to experience this new dimension in the high-performance dynamic experience.

Based on the Neue Klasse architecture, the Vision Driving Experience (VDE) is a demonstrator for the capabilities of the innovative "Heart of Joy" central electronic dynamic control computer network.

The VDE track demonstrator, with up to 18,000 Nm torque and fans giving 1.2 tons downforce for Formula 1 levels of cornering grip. BMW stresses that this is not the new M3, but its influence is sure to be important.

(CONTINUED FROM PAGE 207)

BMW BETS BIG—AND A NEW PHILOSOPHY EMERGES

The answer to the question of whether BMW would stick with combustion, go all electric, or combine the two came in May 2021 when BMW management called a press conference to announce a major new project. Dubbed Neue Klasse, a revival of the iconic name from the previous century, the initiative would be, according to CEO Oliver Zipse, "the biggest future project our company as ever taken on." It would involve the development of a wholly new dedicated electric vehicle platform, a fresh industrial infrastructure to build it, and the adoption of a new philosophy of sustainability, circularity in the use of materials, and zero greenhouse gas emissions.

The choice of the Neue Klasse name, which alongside CSL is one of the most revered in the BMW lexicon, was deliberate: The original Neue Klasse series was responsible for heroically saving BMW when the company was on the rocks in the early 1960s. This miracle turnaround had been made possible only by the models' success with middle-class buyers, BMW managers were keen to point out. Today, BMW clearly no longer needs rescuing, but by choosing the Neue Klasse label, the executives were openly signaling not only the transition to a mass-appeal electric generation—but, just like in the '60s, the start of a new era for the company.

As our timeline on page 8 makes clear, successive announcements and prototype presentations over the following three years served to fill in many of the engineering, electronic, aesthetic, and presentational details of the new wave of models. The new pure-electric architecture would be scalable to suit all sizes and types of vehicle. The new sixth-generation drive system would run at 800 volts, with the attendant advantages of swifter charging, greater efficiency, and easier cabling. New battery chemistries and cylindrical (rather than prismatic) cells would promise much greater energy density for increased range.

Meanwhile, BMW said, a wholly new electronic and software architecture would promise to deliver faster response

ABOVE: Fitting the largely hand-made prototype parts to the interior of the Neue Klasse Vision X in the studio, while (RIGHT) color and trim specialists discuss material choices for the concept.

The first of the Vision series of design studies were the iVision Circular in 2021 (left) and the iVision Dee, shown at the 2023 Las Vegas Consumer Electronics show. They showcased recycled materials and color-change body surfaces respectively.

times, greater reliability, and—crucial for the new range's mission as an ambassador for the digital age—a quantum leap forward in how information is presented to the vehicle occupants. And the unquestionable highlight here would be the use of the entire windshield area as a truly panoramic display for screening anything from technical information to movies, maps, and web content.

After the style shock that had been delivered by the nakedly polarizing XM super-SUV not long beforehand, the clear, fresh, and completely nonaggressive design of the Neue Klasse prototypes came as a welcome relief. Design head Adrian van Hooydonk even said its feel should be "so new that it almost seems as if design has skipped a generation." Of course, there also needs to be a certain familiarity that grounds it as clearly a BMW, summed up by design leader Domagoj Dukec as "you know it's a new car and you've never seen it before, yet it's also strangely familiar."

The Vision Neue Klasse X concept in a CGI-generated environment. Notable is how the twin upright BMW kidney grilles give a very different frontal identity to the lower and sleeker-looking sedan concept.

LEFT: The novel interior architecture of the Vision Neue Klasse X concept vehicle, showing the panoramic screen with the pillar-to-pillar driver-information strip at its base. The central display handles navigation and communication functions, while a head-up display on the windshield provides navigation prompts and alert messages in the driver's line of sight.

BELOW: Flooded with light: Extravagant glass areas give the X concept an exceptionally bright interior, but volume-production versions will adopt a more measured approach that helps minimize weight and the solar gain load on the air conditioning.

The sedan-format Vision Neue Klasse concept brings a more subdued color pallet to its very minimalist interior. The driver-info display band at the base of the screen is clearly visible and takes the place of conventional instrumentation, and familiar switchgear functions are moved to haptic buttons on the steering wheel—despite evidence of a growing consumer backlash against touch-screen operation of key safety-related systems.

Among those familiarizing prompts are the clear three-box silhouette, the return of the shark-nose front in a much more modern interpretation, and the use of light instead of chrome to outline key brand signifiers, such as quad headlights and the kidney grille in digital form.

As with any major BMW model program, internal competitions were held to decide the design for the final production model—and the vote went to the work of Danish designer Anders Thøgersen, the man who had already shaped the three concept models. The natural expectation would be for the final production cars to remain faithful to the concepts in their appearance.

Insiders describe the new-generation cars as "a seismic shift in BMW's style," forsaking the powerful muscularity and great drama of some past models and instead bringing in a sense of relaxation and purity. This is closely aligned with the mid-2020s zeitgeist and taps into the sensitivities of a rising generation of consumers who are not automatically enamored with cars. For them, a car isn't necessarily about performance, handling, or an aggressive image; often, the private experience within the vehicle is equally—or even more—important. Hence the very strong emphasis on the major advances in digital connectivity and information presentation promised by the Neue Klasse's innovative electronic architecture.

Light fantastic: The Vision Neue Klasse sedan (left) and X-designated SUV (right) demonstrating the use of light to project their differentiated brand identities. Early photos of new iX3 test vehicles show proportions and detailing very close to those of the concept model.

WHY THE NEW 3-SERIES IS SO SIGNIFICANT

Even now, before the definitive design has even turned a wheel in public, it is clear that the new Neue Klasse represents a step change on a multitude of levels. Not just in the most obvious areas, such as the switch to electric power or the reinvention of car design as a force for calm and tranquility rather than speed and assertiveness—it goes a good deal deeper than that.

In many ways, the upshift to a new mindset that's gentle, kind, and caring is reflective of a deeper change in BMW itself, a new and more holistic way of thinking that places environmental and sustainability considerations at the top of the agenda. And, in a key point raised by Steve Saxty in his conversations with the top design team, these more holistic attitudes in turn stem from a generational change in BMW's top management. The principal decision-makers are now those born in the 1960s and influenced by the more questioning climate of the '70s and '80s.

Of course, nothing in this new ethos of caring and sharing will be allowed to prevent BMW from remaining a fiercely competitive and profit-driven company whose share price is eagerly tracked by analysts. But the very stable ownership structure provided by the firm's long-standing majority shareholder gives it the luxury of making far-reaching decisions to secure its long-term future, something often hard to achieve in other organizations with a short-term focus on share price and the next quarter's results as the metrics of their success.

As for the vexing question of where this leaves the core 3-Series and the "sheer driving pleasure" leitmotif that it has always inspired, CEO Zipse reassures fans that they have no need to worry. He promised *Autocar* magazine that the Neue Klasse cars will be "more BMW than you have ever experienced before," adding, "We won't shy away from our roots but have freed ourselves from the shackles of the past. We've had the courage to skip a step."

Extrapolating that upshift to the whole company, design director Dukec intimated, in the same interview with Steve Saxty, that this new approach could be transformational for the German manufacturer. "I believe the BMW brand will be a different brand once we've launched the Neue Klasse," he said. "Our promise is that the Neue Klasse will do so many things differently but some things will stay the same."

So, as long as those different things work even better and all the good things stay the same, what's not to like?

BELOW: Sixty years of separation: a 1960s-era 1500 sedan follows the Vision Neue Klasse concept, the likely blueprint for the eighth-generation 3-Series appearing in 2026.

ABOVE: The distinctive tailgate treatment of the Vision Neue Klasse X SUV concept, clearly showing the BMW logo that's not a separate riveted-on component but etched into the sheet metal itself.

ABOUT THE AUTHOR

TONY LEWIN has been a leading auto industry correspondent, commentator, analyst, and author most of his professional life, reporting on the car business; evaluating its products; interviewing its key designers, engineers, and CEOs; and traveling the world in search of exclusive insights.

Fascinated by cars, motorbikes, and engines from a very early age, Lewin began his technical education aboard Hondas and Triumphs in his teens before joining *What Car* magazine in London and becoming its editor for six years. Turning freelance, Lewin then wrote for magazines as varied as *Car, Fast Lane, Performance Car, Autocar, Classic Car*, and *Auto Express*, as well as national and international papers such as the *Guardian*, the *Times*, the *Financial Times, Die Welt*, the *Wall Street Journal*, and the *International Herald Tribune*.

Highlights that still stand out after several decades include driving a Ferrari F40 around Fiorano, taking a Ford GT40 across France, and threading scary classic Lamborghinis from the Miura to the Countach through Italian city streets, twisty mountain passes, and high-speed *autostradas*.

As motoring correspondent for the *European* and several national newspapers, Lewin also began reporting on the auto industry for *Automotive News* and was a key part of the launch of *Automotive News Europe*. Working in parallel with *Financial Times Business*, he edited *FT Automotive Environment Analyst, FT World Automotive Manufacturing*, and, in 1999, the key monthly glossy, *FT Automotive World*, which brought original academic input and penetrating economic analysis to the business world.

Among Lewin's previous BMW books for Motorbooks are *The Complete Book of BMW* (2004), *The BMW Century*, 1st and 2nd editions (2016 and 2022), and *BMW M: 50 Years of the Ultimate Driving Machines* (2021). Non-BMW books for Motorbooks include *How to Design Cars Like a Pro*, 1st and 2nd editions (2003 and 2010), *Smart Thinking: The Little Car That Made It Big* (2004), and *Speed Read, Car Design* (2017). Books for other publishers include six annual editions of the *Car Design Yearbook* (2004–2009), *The A–Z of 21st Century Car Design* (2011), and *London's New Routemaster* (2013), all for Merrells, and *Ricardo: 100 Years of Innovation and Technology* (2015).

Lewin is also an accomplished professional translator, working with publishers, agencies, and the companies themselves. Among his translated works are, from German, *Junkyard: Behind the Gates at California's Secretive European-Car Salvage Yard*, for Motorbooks, and, from French, *Design Between the Lines*, by Patrick le Quément, for Merrells.

Lewin lives in southeast England, now drives an electric BMW, and enjoys exploring, hiking, and riding bikes—though now strictly the human-powered variety. As for cars, he insists he's not finished yet.

ACKNOWLEDGMENTS

It takes much more than a single individual to compile a book like *BMW 3-Series*. So while I may have carried overall responsibility for weaving the many different strands together, without the people to provide those countless threads, the tapestry would not exist at all. My gratitude goes out to all.

Great floods of facts and figures from BMW's central corporate archives; the steady flow of insights from company insiders and intriguing revelations from people in power; perceptive observations from the workshop floor and from owners and clubs—punctuated by the occasional absolute gem from an unexpected source—all were vital streams in the mix.

In this age of misinformation, distortions, and outright lies, trusted sources are doubly essential to any accurate record of events, and here BMW itself proved invaluable from the get-go. The Group's PR operations in Germany, the UK, and the US were exemplary in their supply of everything from intricate technical data to broad-brush sales statistics (so any mistakes on the numerical front are mine), and I'm deeply grateful to them for granting me access to top executives for cross-questioning. Special thanks, as well, to BMW Classic and the Group Media Archive, who went the extra mile to find images of rare models that were missing from our own collection.

Closer to home, my colleague and friend Steve Saxty, fresh from completing his three-volume magnum opus on BMW design, *BMW Behind the Scenes*, kindly allowed me use of selected designer sketches from his book and even provided me with exclusive never-before-published historic design drawings. Miraculously, seasoned auto journalist turned industry analyst Karl Ludvigsen, a veteran of the 3-Series' 1975 unveiling, was able to track down his original launch report for *Car & Driver*, as well as his contemporary photography, so a big shout-out to him—thank you, Karl. While on the subject of photos, I am indebted to James Mann, who not only took scores of the images you see on these pages but was also able to decode my list of target models, track down people who owned those cars, and persuade them to polish up their prized wheels for photo shoots all across southern England. Those proud owners deserve a name-check, too: Malcolm Field, Steve and Charlotte Harvey, Julian Pickering, Alex Pulford, Nino Radojcin, Will Smith, and Ryan Thomson. Steve even flew back from New York so that he could drive his M3 straight to the shoot.

BMW Car Club GB was instrumental in connecting James with the right cars, and the editors of their *Straight Six* monthly magazine, Jeff Heywood and Will Beaumont, went the extra mile to guide me through the club's ethos and facilitate a survey of members who'd had 3-Series models—which turned out to be most of them.

Anders Bilidt of auctioneers RM Sotheby's is a particularly enlightened owner, and a serial buyer and seller of BMWs. What he doesn't know about BMWs isn't worth knowing, but even more impressive is his unquenchable thirst for more and more examples of the marque and the sheer depth of his knowledge about everything BMW. Thank you, Anders, for your valuable insights both from the auction room and from the rarefied ambience of your personal collection.

Equally insightful are the various BMW specialists dotted around the UK. Special thanks go to Dan Norris of Munich Legends and Barry Sheward of BMR Performance, who were particularly helpful. Between them, they lifted the lid on what keeps older BMWs going, what can unfairly condemn newer ones, and why I should never trust a cam chain again—and also for luring me away from my E46 fixation and seeing the E36 afresh.

And, of course, I cannot possibly forget the patience and tolerance of all who surround me here at BMW Book Central. For putting up with everything from blockade-height stacks of reference books to teetering towers of papers and notebooks, crack-of-dawn departures to catch key interviewees, and the inevitable mood swings depending on how things were going.

Finally, and maybe most importantly of all, I'd like to thank the hundreds and thousands of men and women, designers, planners, engineers and assemblers, fixers and fettlers in the BMW universe. Thank you for producing such great, inspirational cars. Your energy, ingenuity, and enthusiasm shine through in every product you produce—which not only makes every model a thrill to drive but has also made this book such a pleasure and privilege to write.

Tony Lewin
April 2025

INDEX

D

E

F

G

H

I

J

K

L

M

N

O

P

A = all; B = bottom; L = left; M = middle; R = right; T = top

dpa picture alliance / Alamy Stock Photo: 193

Drive Images / Alamy Stock Photo: 116TL

James Mann: 5, 37A, 52A, 53, 54A, 55B, 62TR, 62BR, 64, 66A, 68, 70T, 71B, 74, 78B, 82, 84L, 88, 89L, 96R, 97T, 101A, 103, 104, 109B, 110, 111A, 113A, 115A, 124, 137L, 194

Karl Ludvigsen: 45, 48A, 49B

Magic Car Pics: 62BL, 70B, 81, 179B

Matthew Richardson / Alamy Stock Photo: 86B

Phillip Laws / Alamy Stock Photo: 176

Roman Belogorodov / Alamy Stock Photo: 109T

Ted Soqui / Getty Images: 135B

Tony Lewin: 180, 181, 182T

BMW i VISION Dee

BMW i VISION Dee